D0215709

STRUCTURE AND BONDING

Volume 28

With 25 Figures and 15 Tables

Springer-Verlag
Berlin Heidelberg New York 1976

ISBN 3-540-07753-7 Springer-Verlag Berlin Heidelberg New York
ISBN 0-387-07753-7 Springer-Verlag New York Heidelberg Berlin

Library of Congress Catalog Card Number 67-11280

Printed in Germany

Typesetting: R. & J. Blank, München. Printing and bookbinding: Brühlsche Universitätsdruckerei, Gießen

Contents

SPRINGER-VERLAG

D-6900 Heidelberg 1
P. O. Box 105280
Telephone (06221) 487·1
Telex 04-61723

D-1000 Berlin 33
Heidelberger Platz 3
Telephone (030) 822001
Telex 01-83319

SPRINGER-VERLAG
NEW YORK INC.

175, Fifth Avenue
New York, N.Y. 10010
Telephone 673-2660

Reversible Oxygenation

R. W. Erskine and B. O. Field

The Inorganic Chemistry Laboratory, The City University, London, ECIV 4PB

Table of Contents

Abbreviations

acacen = N,N′-ethylene bis (acetylacetoneiminato);
Bae = N,N′-ethylene bis (acetylacetone-iminato) dianion;
bzacen = N,N′-ethylene bis (benzoylacetoneiminato);
t-Bsalten = N,N′-(1,1,2,2-tetramethyl) ethylene bis (3-tertbutylsalicylideniminato);
1-n-BuIm = 1-n-Butylimidazole;
4-t-BuIm = 4-t-Butylimidazole;
DMG = Bis(dimethylgyloximate);
DGENTA = N,N,N‴,N‴-Diglycylethylenediaminetetraacetic acid;
DMF = Dimethylformamide;
Diphos = Cis-($(C_6H_5)_2$P $CH_2 \cdot CH_2$ P$(C_6H_5)_2$);
EFG = Electric field gradient;
EH = Extended Hückel;
ESR = Electron spin resonance;
Hb = Hemoglobin;
H_2Tam PP = Meso-tetra (*o*-aminophenyl) porphyrin;
H_2Tpiv PP = Meso-tetra (*o*-pivalamidophenyl) porphyrin;
IR = Infra red;
L-Dampa = L-2,3-diaminopropionate;
1-MeIm = 1-Methylimidazole;
MO = Molecular orbital;
Mb = Myoglobin;
NMR = Nuclear magnetic resonance;
pip = piperidine;
py = pyridine;
(2 = phos) = Cis-($(C_6H_5)_2$P CH = CH P$(C_6H_5)_2$);
Salen = N,N′-ethylene bis (salicylaldiminato);
Salen C_2H_4 py = α,α′-(2-(2′-pyridyl)-ethyl) ethylene bis-(salicylideniminato);
Salpn = N,N′-trimethylene bis (salicylaldiminato);
Saloph = N,N′-*o*-phenylene bis (salicylidene-iminato) dianion;
Sal Me DPT = Bis (salicylidene-γ-iminopropyl) methylamine;
SCC = self-consistent-charge;
SDTMA = N,N-bis(2-aminoethyl) glycine;
THF = Tetrahydrofuran;
1-trityl Im = 1-triphenylmethylimidazole;
Trien = Triethylenetetramine;
TPP = Tetraphenylporphyrin;
UDTMA = Diethylenetriamine-N-acetic acid.

I. Introduction

Much work has been carried out recently on some new synthetic porphyrins, 'picket-fence' porphyrins being especially worthy of note, and these have provided excellent models for the naturally occurring oxygen carrying molecules myoglobin and hemoglobin. These synthetic complexes have been amenable to structural work, as well as other physical and chemical studies, and have thus resolved many of the controversies which have concerned the orientation of the dioxygen ligand in oxyhemoglobin, as well as the factors affecting reversibility.

This review is particularly concerned with the bonding modes of the dioxygen ligand, and the factors affecting which of the various possible orientations it assumes in any particular complex. We shall therefore consider all previous work, from the earliest work on synthetic oxygen carriers, right up to the most recent studies on 'picket-fence' and other synthetic porphyrins. All work concerned with an irreversible system will illustrate some principle appertaining to reversibility. On this basis we shall then attempt to provide a unified rationale for:

A. the orientation of the dioxygen ligand in dioxygen complexes, and
B. the factors affecting reversibility.

The form of this review, which is presented in an historical context, attempts to illustrate the evolution of ideas and opinions on the subject matter up to the present day. To help achieve this objective, Section III has been restricted mainly to older studies; however, much of the work covered in this section is clarified or challenged by work discussed in later sections so should not be taken in isolation. The work on naturally occurring and synthetic porphyrins is included together in Section V, the older studies being discussed in the introduction to the section.

The main factor which distinguishes the more recent work is the prevalence of definitive structural data (i.e. crystallographic analysis), which is of primary importance in achieving the objective (*A*), above. The reactions of coordinated dioxygen are not discussed in any depth here (except in the case of bimolecular termination which results in the formation of bridging dioxygen complexes); this subject clearly warrants a separate review. We have attempted to present a succinct review, illustrating principles wherever possible, rather than simply providing a catalogue of data. To avoid ambiguity it is important to stress that the terms,

'molecular oxygen', the 'oxygen molecule', and 'dioxygen' all refer to O_2, whereas 'oxygen' refers to O.

'Dioxygen' is generally used to describe coordinated O_2.

'Oxygen-carrier' refers to an O_2-carrying molecule.

II. The Dioxygen Ligand

A. The Oxygen Molecule

According to molecular orbital theory the valence *p*-orbitals of the oxygen atoms combine to form three bonding and three antibonding orbitals in molecular oxygen, as shown in Fig. 1.

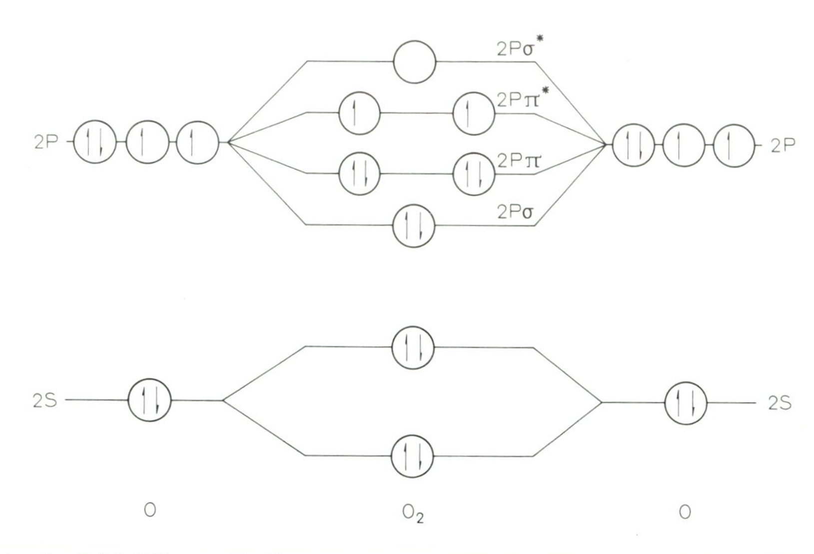

Fig. 1. Molecular Orbital Diagram for O_2

Molecular orbital theory successfully predicts the observed paramagnetism of the oxygen molecule, and also predicts that if electrons are added or removed (initially to or from the $2\,p\pi^*$ orbitals), the bond order will be lowered or raised accordingly. Hence the dioxygenyl cation O_2^-, molecular oxygen O_2, superoxide O_2^-, and peroxide O_2^{2-} are predicted to have bond orders of 2.5, 2, 1.5 and 1 respectively (*1*). Having correlated the bond order and MO distribution in this way, later arguments will depend heavily on the concept that when dioxygen is coordinated, there is partial exchange of electron density via appropriate orbitals on the dioxygen ligand and metal concerned, which in turn produces a corresponding change in the O–O bond order.

B. The Reactivity of the Dioxygen Ligand

In principle, a total of 4 electrons may be progressively added to the oxygen molecule, eventually resulting in zero bond order – hence the ultimate reduction product is oxide.

The redox potential at pH = 0.0 for the reaction

$$O_2 + 4\,H^+ + 4\,e^- \rightarrow 2\,H_2O$$

is 1.23 V (*2*), indicating that molecular oxygen is a strong oxidizing agent. Certain enzymatic oxidation reactions (*4,5*) are thought to involve a synchronous four-electron process. In the synthetic dioxygen complexes that have been studied to date, the formal reduction of the dioxygen ligand never proceeds beyond peroxide (O_2^{2-}) formation; that is, the net donation to the co-ordinated dioxygen never exceeds two electrons, which gives rise to an oxygen-oxygen single bond.

A full discussion of the reactions of molecular oxygen involved in dioxygen adducts would require a separate review. Moreover, the reactions of co-ordinated molecular oxygen are discussed elsewhere (*1–8*); dioxygen adducts of biological systems (*7*) and synthetic group VIII metal complexes (*1,3*) being of particular interest.

C. Classification of Bonding Types for the Dioxygen Ligand

In order to systematise the literature, the following classification is used with respect to the metal dioxygen bonding orientations found in oxygen carriers.

Type 1. Bridging adducts (see Fig. 2), where the dioxygen ligand is bridged between two metal atoms intermolecularly. Here, the metal dioxygen ratio is invariably 2 : 1.

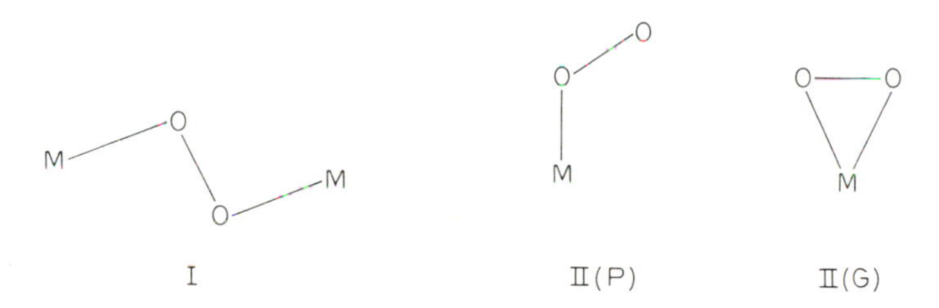

Fig. 2. Classification of bonding modes for Dioxygen

Type II. Non-bridging adducts (see Fig. 2), where the dioxygen ligand is bonded to just one metal. Here, the metal: dioxygen ratio is generally 1 : 1. (However, in principle a square planar complex, for instance, may be capable of coordinating two dioxygen ligands providing the electronic factors do not prohibit this.) Within this classification we find that there are two possible orientations II(*P*) and II(*G*), the bonding involved is discussed later, see in particular section V(*B*). The (*P*) and (*G*) labels refer to *Pauling* (*30, 31*) and *Griffith* (*32*) who originally proposed these structures for the bonding of dioxygen in HbO_2.

III. Older Studies

A. *Early work on Cobalt(II) Salen and Related Chelates*

The first example of a synthetic reversible oxygen adduct was prepared by *Pfeiffer et al.* (*9*) in 1933, who observed that red-brown crystals of cobalt(II) salen darkened on exposure to air. *Tsumaki* (*10*) later confirmed that this colour change was due to reversible oxygenation. *Calvin et al.* (*11–16*) studied a variety of related complexes and found that two types of chelate reversibly bind O_2 in the solid state and in certain solvating solvents (*17*), such as quinoline. The unsubstituted chelates are given by *A* and *B* in Fig. 3. In aprotic media there is no possibility to increase the transfer of electrons to the dioxygen bridge via hydrogen bridges or to form H_2O_2, therefore, the lower the donor and acceptor capabilities of the solvent used, when oxygenating Co(II) salen in solution, the more stable will be the dioxygen adduct towards irreversible oxygenation (*25*). Both chelates can exist in several crystalline modifications and in its active states type *A* forms dioxygen complexes with a Co : O_2 ratio of 2 : 1. All forms of the Co(II) salen complex contain the same molecules since they give identical infra-red in chloroform solution (*18*). Analogues of these chelates have been prepared (*12, 15*) using Mn(II), Fe(II), Ni(II) and Cu(II) as the central metal ion, but none of them exhibit reversible oxygenation, the metals appear to be irreversibly oxidized to the corresponding M(III) species.

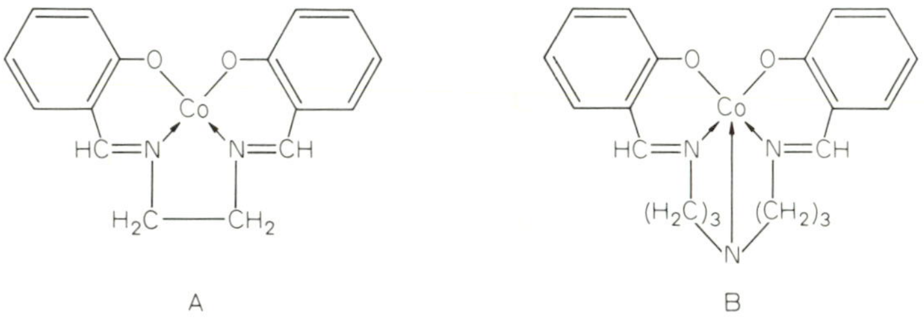

Fig. 3. A is abbreviated as Co(II) Salen

Calvin et al. investigated the oxygen-carrying chelate Co(II) salen, including the study of rates of oxygenation (*13*), the recycling properties of the dioxygen complex (*14*), the magnetic properties (*15*), equilibria studies (*16*) and X-ray studies (*12*). Co(II) salen was consecutively oxygenated and deoxygenated, at room temperature and 80–100 °C respectively; it was found to deteriorate to 70% of its original activity after 300 cycles (*14, 19*). Other related chelates also display deterioration; the 3-fluoro Co(II) salen derivative deteriorates to 60% of its initial activity after 1500 cycles (*12, 14*).

This deterioration is attributed mainly to irreversible oxidation (*12, 14, 17*), but other factors may be significant, for example strain in the crystal on oxygenation may eventually lead to an inactive modification (*17*). Co(II) salen is paramagnetic (low spin d^7 electron configuration), and its (2 : 1) dioxygen adduct is diamagnetic, suggesting the formation of a peroxo linkage (*2*).

Also, X-ray work carried out on Co(II) salen by *Calvin et al.* (*12*) shows that 'salen' is bonded to the cobalt atom in a square planar configuration. Measurement of the area covered by one Co(II) salen molecule was also used to establish its planarity (*20*). Planar Co(II) complexes have a characteristic *d–d* transition in the visible range at 1200 mμ (*21*), which has been observed in Co(II) salen (*22*), and some workers (*23, 24*) have concluded that low spin Co(II) complexes are generally planar. The planarity is thought to be an important factor in the oxygen-carrying ability of Co(II) salen (*19, 25*), and, on the basis of infra-red evidence, it was concluded (*18, 22, 25*) that oxygenation has negligible influence on the planarity of the chelate. Solid Co(II) salen changes colour from brown to black on oxygenation (*17*), and in MeOH solution the dioxygen adduct gives a charge transfer absorption (*22*) at 665 mμ which appears to be polarized perpendicularly to the plane of the chelate. The suggestion (*26–28, 17*) that deoxygenated Co(II) salen exists in a dimeric form containing an aquo bridge, and that this is essential for the complex to function as a synthetic oxygen carrier, has not been confirmed by infra-red data (*29*). Moreover, Co(II) salen was dehydrated by refluxing in pyridine under nitrogen and then, after recrystallization and removal of solvated pyridine, was still found to be an active oxygen carrier (*29*).

More recent studies on Co(II) salen and related derivatives are discussed in Section IV(*A*), and these will clarify and extend the older work just considered, particularly from a structural point of view.

B. Amino Acid and Dipeptide Complexes of Cobalt(II)

Cobalt(II) forms many complexes which can exhibit oxygen-carrying properties (*2, 19*). Reversible oxygen uptake in solutions of cobalt(II)-histidine (*33–36*), and cobalt (II) in the presence of α-amino acids and peptides (*37–39*) has been known for some time. The reaction of cobalt(II) with dipeptide was first observed in enzymic studies involving glycyl-glycine (*40*).

The high spin octahedral (*45*) complex bis (histidinato) cobalt(II) reacts rapidly (*41*) with molecular oxygen in aqueous media to give a deep brown solution containing the diamagnetic (*42*), binuclear (Co : O_2 = 2 : 1) oxygenated complex (*33*), which has been isolated (*36*). Accompanying this rapid reversible dioxygen uptake is a slow irreversible uptake of more O_2 which leads eventually to the formation of dark pink cobalt(III) complexes (*41*), thought (*35*) to contain hydroxo and not peroxo groups. A polarographic study (*35*) indicates that an intermediate exists in the reversible reaction with bis(histidinato) cobalt(II), and the kinetics (*41*) are consistent with the following reaction scheme:

$$\mathrm{Co(h)_2 + O_2 \rightarrow Co(h)_2O_2\ ;\quad Co(h)_2 + Co(h)_2O_2 \rightarrow [Co(h)_2]_2 \cdot O_2}$$

(where h = histidine).

In strongly basic solution this reaction involves the formation of an hydroxo bridge in addition to the dioxygen bridge (*43*). The redox potential for the process

$$M(II) + O_2 \rightarrow M(III)\ (M = Fe, Co, Ni, Cu)$$

is in the order Fe $<$ Co $<$ Ni $<$ Cu(2), and this might explain, on the one hand, the inactivity of nickel(II) and copper(II) histidine complexes towards molecular oxygen, and on the other hand, the irreversible oxidation of iron(II) histidine to iron(III) by dioxygen. The oxygenation of the Fe(II), Co(II) and Mn(II) chelates of tetraethylenepentamine was studied (*51*) and, again, reversible oxygenation was observed only with the cobalt chelate. However, the nickel complex of tetrapeptide tetraglycine is readily oxidized by molecular oxygen under mild conditions (*44*); the nickel is thought to be activated by peptide coordination. An aqueous solution containing 1, 2 diaminoethane and Co(II) ions was oxidized by molecular oxygen to give a μ-peroxo-μ-hydroxo complex (*46*). *Gillard* and *Spencer* (*47*) have measured the oxygen uptake of nineteen cobalt(II) dipeptide complexes in alkaline solutions. They concluded (*47*) that a minimum of three (*48*) nitrogen donors must be chelated to the cobalt for the resulting complex to display oxygenation; the donating power of the ligands is correlated with the ability to stabilize the $Co(II) \cdot O_2$ moitey. Although the oxygenation of cobalt(II) dipeptides is reversible, there are found (*49, 50*) to be other species in solution arising from a complex series of side equilibria, which eventually results in the formation of mononuclear cobalt(III) chelates. Proton balance studies (*48*) show that the binuclear oxygenated cobalt(II) complexes of ethylenediamine, histamine, and glycinamide contain hydroxo as well as dioxygen bridges; the hydroxo bridge is thought to stabilize the adduct. The absence of hydroxo bridging with other ligands, such as diamino-propionic acid, histidine, histidinamide, and histidylglycine is attributed (*48*) to steric effects.

C. *Complexes of Dimethylglyoxime and Related Ligands*

It has been noted (*2*) that many oxygen-carrying cobalt(II) complexes involve ligands that provide four donor atoms lying in the same plane as the cobalt atoms, such as 'salen', porphyrin and dimethylglyoxime, DMG, and related ligands.

It has been observed (*52*) that $Fe(DMG)_2$ readily takes up dioxygen in the presence of ligands such as pyridine, ammonia, histidine or imidazole; bubbling nitrogen through the solution reverses the process

$$Fe(II)\,(DMG)_2\ (base)_2 + O_2 \rightarrow Fe(II)\,(DMG)_2\ (Base)\,O_2 + base.$$

The reversible oxygenation of $Fe(DMG)_2\ (B)_2$ in aqueous dioxane solution, as monitored by reversible changes in the visible absorption spectrum and manometric oxygen evolution, has been reported (*53*), but in a more recent study (*54*) it was not possible to confirm the earlier findings. However it was established that there was a slow irreversible autoxidation to Fe(III), followed by reduction via imidazole or DMG to Fe(II) when O_2 is purged from the solution. Reversible dioxygen uptake has been reported (*55*) for strongly alkaline solutions of nickel(II) $(DMG)_2$, and cobalt(II) $(DMG)_2$ has been studied (*56*) as a possible model for the oxygen carrier vitamin B_{12r} (*57, 58*).

Cobalt(II) $(DMG)_2$ can form monomeric and dimeric dioxygen adducts in the solid state, whereas in solution it only forms weakly bonded dimeric adducts (*56*); these species are readily oxidized to cobalt(III), especially when traces of water are present. The initial dioxygen adducts are diamagnetic peroxo-bridged complexes which can be isolated as solids. Solutions of these adducts in solvents such as dichloromethane, benzene and acetone, (which are coordinating solvents), gradually turn brown with the evolution of oxygen, and the formation of a superoxo-bridged species:

$$
\begin{aligned}
&2\,(\text{base})\,(\text{DMG})_2\,\text{Co–O}_2\text{–Co(DMG)}_2\,(\text{base}) \\
&\rightarrow [(\text{base})\,(\text{DMG})_2\,\text{Co–O}_2\text{–Co(DMG)}_2\,(\text{base})]^+ \\
&+ 2\,\text{Co(DMG)}_2\,(\text{base}) + \tfrac{1}{2}\,\text{O}_2 + \text{H}_2\text{O}\,.
\end{aligned}
$$

The ESR spectrum of the superoxo-bridged species suggests that it undergoes further change to form the mononuclear peroxo radical species (base)–$Co(DMG)_2O_2^{\cdot}$ (*56*), and the ESR spectrum of Vitamin B_{12r} (*58*), on reaction with dioxygen, is very similar indicating the formation of a peroxo radical. It is probable (*56*) that solutions of cobalt(II) chelates in, say, CH_2Cl_2 react with molecular oxygen in two stages to give peroxo-bridged complexes:

$$
\begin{aligned}
&\text{Co(II)} + \text{O}_2 \rightarrow \text{Co}\cdot\text{O}_2^{\cdot} \\
&\text{Co}\cdot\text{O}_2^{\cdot} + \text{Co(II)} \rightarrow \text{Co–O}_2\text{–Co}.
\end{aligned}
$$

It is possible that steric factors prevent the formation of the bridging peroxide species in the case of vitamin B_{12r} (*58*), however, other cases have been reported (*59*) in which oxygenation proceeds no further than the formation of mononuclear peroxo radicals (i.e. first stage above).

D. Air oxidation of Cobalt(II) Ammine complexes

It has been known for some time (*60–63*) that aqueous ammonia solutions containing cobalt(II) salts turn brown on exposure to air. Thermodynamic (*64*) and kinetic (*65, 66*) studies on cobalt(II) ammines have therefore been carried out in inert atmosphere.

The diamagnetic salt of a binuclear complex ion,

$$[(NH_3)_5Co{-}O_2{-}Co(NH_3)_5]^{4+}, \qquad \text{(A)}$$

can be isolated from a solution of Co(II) and NH_3 which has been saturated with molecular oxygen. The solid releases oxygen at 100 °C under vacuum (*67*). In aqueous solution a variety of treatments including nitrogen purging and acidification can be used to remove the dioxygen ligand, and this enables successive oxygenation and deoxygenation cycles to be carried out. Experiments (*67*) involving ^{18}O have shown that the bridging dioxygen ligand in (*A*) originates entirely from the molecular oxygen used in its preparation. Treatment of (*A*) in aqueous solution with hydrogen peroxide gives the green paramagnetic species;

$$[(NH_3)_5Co{-}O_2{-}Co(NH_3)_5]^{5+}, \qquad (B)$$

an ESR study (*68*) of this complex shows that the cobalt atoms are in identical environments, and it seems likely that the odd electron is delocalized over the two cobalt atoms and dioxygen bridge. If this electron orbital was heavily localized on the dioxygen bridge, then we would expect a superoxide species and in fact X-ray studies (*69, 70*) carried out on (*B*) show that the Co–O–O–Co moiety is planar, with the bridging dioxygen group skewed to the Co–Co axis and an O–O bond length of 0.131 nm, which is indicative of a superoxide species. Moreover, an X-ray study (*71, 72*) of the diamagnetic species (*A*) shows that the Co–O–O–Co moiety is non-planar (torsional angle of 146° about the O–O bond) and that the O–O bond length is 0.147 nm; both observations indicate the presence of a peroxo linkage. The cobalt atoms are in approximately octahedral environments consisting of five NH_3 groups and an oxygen atom. In oxygen free aqueous solutions of Co(II) and NH_3 (in high concentration) the most important ions present are the six coordinate hexammine and aquated pentammine Co(II) species:

$$Co(NH_3)_6^{2+} + H_2O \rightarrow Co(NH_3)_5(H_2O)^{2+} + NH_3$$

Both are active towards molecular oxygen with which they react rapidly through second order (*67, 41*) rate processes (*73*):

$$2\,Co(NH_3)_5(H_2O)^{2+} + O_2 \rightleftharpoons (A) + 2\,H_2O$$
$$2\,Co(NH_3)_6^{2+} + O_2 \rightleftharpoons (A) + 2\,NH_3$$

There is tentative evidence for the lower reactivity of the hexaammine species in high ammonia concentration. The tetraammine (diaquated) and lower ammine Co(II) species do not pick up O_2 over relatively long periods (*73*).

Complex (*B*) can be reduced in acid aqueous medium by Fe^{2+} to complex (*A*), the process results in the evolution of molecular oxygen (*74*), this being ascribed to the decomposition of (*A*). In analogous reductions using Cr^{2+}, V^{2+} and Eu^{2+}, a common intermediate was detected (*75*), which has been assigned the structure

$$[(NH_3)_5Co{-}O(OH){-}Co(NH_3)_5]^{5+}$$

and which decomposes at a rate slower than that observed (*73*) for (*A*) in alkaline solution. Protonation in the former intermediate species is thought (*73*) to be responsible for the difference in rates of decomposition.

IV. More Recent, Definitive Studies

Further to section III(*C*), we shall see that type I dioxygen complexes can be formed in solution via intermediates of a type II(*P*) structure. Consequently, sections IV(*A*) and IV(*B*) will tend to overlap in places. The dioxygen adducts of some new synthetic metal porphyrins belong to the II(*P*) classification. However, a discussion of the dioxygen adducts of naturally occurring and synthetic metal porphyrins is postponed until section V. Previous reviews on dioxygen complexes (*1, 19, 25, 76–79*) cover the work up to and through 1972, and will not be totally incorporated within this present review, in the sense that we intend to provide a clarification of the concepts and principles appertaining to reversible oxygenation and not an exhaustive catalogue of data, for which we cite the original papers and reviews. We cover the literature through 1975.

A. Type I Dioxygen Complexes

A large number of type I dioxygen adducts are reported in the literature (*25, 76, 80*). The vast majority of metal complexes which take up dioxygen to form type I adducts contain cobalt(II) as the metal ion. *Ochiai* (*81*) considered the hypothetical reactions:

$$M(II) + O_2 \rightleftharpoons M \cdot O_2 \qquad (A)$$

$$2\,M(II) + O_2 \rightleftharpoons M{-}O_2{-}M \qquad (B)$$

where M = metal

Having estimated the Gibbs free energy values for reactions (*A*) and (*B*), using thermodynamic data obtained in aqueous media, it was concluded (*81*) that the formation of the μ-peroxo ($M : O_2 = 2 : 1$) type complex is favoured over that of the superoxo (1 : 1) type complex. Also it was deduced (*81*) that the M(II) species that could be expected to form at least some reversible dioxygen adducts are, in the case of reaction (*B*), Co(II), Fe(II) and Mn(II), and in the case of reaction (*A*), Co(II), Fe(II), and less favourably, Ti(II), Cr(II) and Mn(II). However, the stability of these dioxygen adducts will depend critically on kinetic factors; in aqueous media, the only moities considered (*81*) to be stable against substitution of the dioxygen ligand, leading to oxidation of the metal, are $Co \cdot O_2$, in reaction (*A*), and $Co{-}O_2{-}Co$, in reaction (*B*). Therefore, by considering both the thermodynamics and kinetics of dioxygen uptake *Ochiai* (*81*) has provided a rationale for the observed preponderance of binuclear (2 : 1) cobalt dioxygen adducts; and he also points out that the process

$$Co(II) \rightarrow Co(III) + e^-$$

is thermodynamically unfavourable unless a strong coordinating ligand such as CN^- is present, which can stabilize Co(III) and thus prevent a peroxo bridge being oxidized to superoxo. An ESR study has been carried out (*68*) on a variety of cobalt(II) amine complexes in aqueous media. An ESR signal was produced (*68*) when O_2 was bubbled through the Co(II) amine aqueous solutions which was very similar to that found for monomeric dioxygen adducts of cobalt(II) Schiff bases, and which corresponds to the interaction of an unpaired electron with a single ^{59}Co nucleus of spin 7/2, and indicates the presence of a peroxy radical (*82*). The ESR signal decayed with time, accompanied by an increase in optical density of the solution in the region 300–400 mμ. This optical absorption corresponds to that found (*83*) for binuclear dioxygen cobalt complexes; therefore, the disappearance of the ESR signal is attributed (*68*) to bimolecular termination between the peroxy radical and the paramagnetic chelated cobalt atom, thus:

$$(\mathrm{Chel})\mathrm{Co{-}O_2^{\cdot}} + \mathrm{Co(Chel)} \rightarrow (\mathrm{Chel})\mathrm{Co{-}O_2{-}Co(Chel)}$$

where Chel = chelate

These conclusions are supported by another ESR study (*56*) made on some oxygenated cobaltoximes(II), which showed that the dioxygen species finally formed is best characterized as a peroxo-bridged complex; these being reversibly formed binuclear complexes.

It has been noted (*84*) that μ-dioxygen (type I) complexes formed in solution often contain a μ-hydroxo bridge, in addition to the μ-dioxygen bridge, provided that each cobalt atom has a vacant coordination site. Fig. 4 illustrates the dibridged M–(O_2, OH)–M moitey

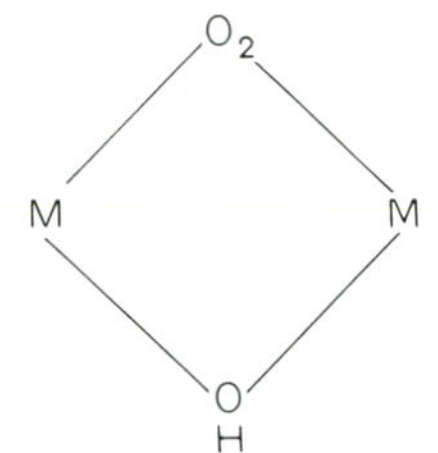

Fig. 4. μ-dioxygen, μ-hydroxo complex

A binuclear peroxo cobalt(III) complex, formed by the oxygenation of solutions containing two moles of L-dampa per mole of Co(II), becomes dibridged at pH > 9 with the formation of a μ-hydroxo bridge, in addition to the peroxo bridge already present (*85*); this was established from the circular dichroism patterns and absorption spectra for the monobridged and dibridged species. The monobridged peroxocobalt(III) complex is characterized (*85*) by its absorption spectrum which contains two maxima within the range 310 to 450 nm, whereas the dibridged (O_2, OH) cobalt(III) complex shows only one band in this range; these assignments were supported (*85*) by titration and circular dichroism experiments. It was also found that on treatment with H_2O_2 the monobridged peroxo cobalt(III) complexes were dissociated into mononuclear Co(III) species. A study has been made of the oxygenation of a variety of polyamine cobalt(II) chelates in aqueous solution (*86*). Rapid reversible dioxygen uptake was accompanied by slow irreversible changes which were attributed (*86*) to the formation of a dibridged species:

$$2\,CoL_3^{2+} + O_2 \underset{rapid}{\rightleftharpoons} [L_3Co{-}O_2{-}CoL_3]^{4+}$$

$$[L_3Co{-}O_2{-}CoL_3]^{4+} \xrightarrow[fast]{} (L_2Co\begin{matrix} O_2 \\ \diagup\quad\diagdown \\ \diagdown\quad\diagup \\ L \end{matrix}CoL_2)^{4+} + L$$

(the initial equilibrium probably proceeding via the formation of a mononuclear peroxy radical), and the formation of the dibridged species was seen (*86*) to be analogous to the process

$$(NH_3)_5Co{-}O_2{-}Co(NH_3)_5^{4+} \rightarrow (NH_3)_4Co{-}(O_2, NH_2){-}Co(NH_3)_4^{3+} + NH_4^+$$

which proceeds slowly at 35 °C in aqueous ammonia. The formation constants for cobalt(II) amine and polyamine complexes have been obtained (*87*) in O_2 free solution. This data was used in a thermodynamic and kinetic study (*86*) of the oxygenation process for certain cobalt(II) polyamines including diethylenetriamine, histamine and ethylenediamine. The kinetic data demonstrated that a two step process was involved in the formation of the reversible dioxygen adducts, as illustrated in principle by

$$M + O_2 \rightleftharpoons M\cdot O_2 \overset{M}{\rightleftharpoons} M\cdot O_2\cdot M$$

These workers noted (*86*) that complexes unreactive to dioxygen have ligand systems such as carboxyl and heterocyclic groupings which tend to withdraw electron density from the metal centre, this then decreases the metal to dioxygen electron donation which is apparently required for the formation of the type I dioxygen adduct. Moreover, the requirement, discussed in section III(*B*), that a minimum of three nitrogen donor atoms are required for the oxygenation of cobalt(II) chelates, can now be rationalized in terms of the greater ability of nitrogen to donate electron density compared with oxygen. Hence, a switch from cobalt-peptide O to cobalt-peptide *N* bonding, which would probably result (*86*) from the ionization of the peptide hydrogen in solution, should produce an increase in sensitivity towards O_2. *Fallab* showed (*88*) that oxygenation of an aqueous solution of cobalt(II) and triethylenetriamine (tren) produces a brown complex with the composition 2 Co : 2 trien: 1 O_2, which releases molecular oxygen at pH 2. This complex was later (*89*) shown to be the dibridged complex (trien) Co–(O_2, OH)–Co(trien)$^{3+}$, (*X*). A more detailed study (*90*) of the kinetics of formation for the complex (*X*) over the pH range 7.5 to 12.6 has shown that the following reaction scheme is likely.

$$Co(trien)(H_2O)_2^{2+} + O_2 \rightarrow (trien)\underset{OH}{Co}{-}O_2{-}\underset{OH_2}{Co}(trien)^{3+} \quad \searrow{-OH_2}$$

$$+H^+ \uparrow\downarrow \qquad\qquad +H^+ \uparrow\downarrow \qquad\qquad (X)$$

$$Co(trien)(OH)(H_2O)^+ + O_2 \rightarrow (trien)\underset{OH}{Co}{-}O_2{-}\underset{OH}{Co}(trien)^{2+} \quad \nearrow{-OH^-}$$

The monobridged intermediates are both formed by second order rate processes, they then rearrange to produce the dibridged product (*X*) through first order processes. The structures were assigned (*90*) on the basis of the extent and rate of pH changes accompanying these stages. The dioxygen cannot be removed from (*X*) by purging with nitrogen, which is the method generally used in the deoxygenation of monobridged dioxygen adducts in solution. However, the O_2 can be recovered by acidifying the solution since this leads (*90*) to the breakdown of (*X*), thus:

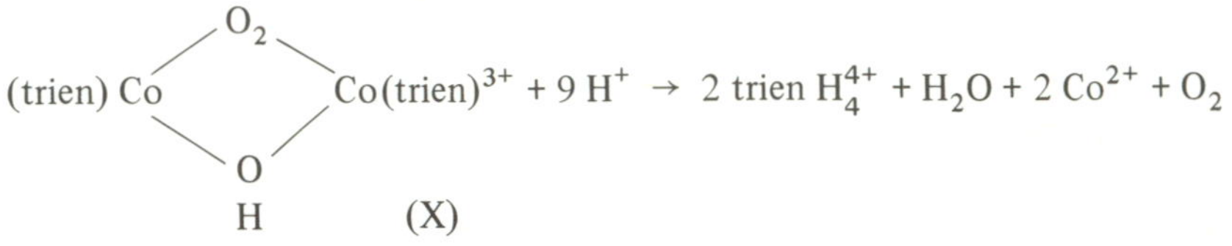

$$(\text{trien})\,\text{Co}\underset{\underset{\text{H}}{\text{O}}}{\overset{\text{O}_2}{\diamond}}\text{Co(trien)}^{3+} + 9\,\text{H}^+ \rightarrow 2\ \text{trien}\ \text{H}_4^{4+} + \text{H}_2\text{O} + 2\,\text{Co}^{2+} + \text{O}_2$$

(X)

A potentiometric study (*91*) of the same system in the pH range 4 to 8 comes to similar conclusions concerning the formation of (*X*). Also, the μ-hydroxo bridge in (*X*) is believed to somehow "lock" the dioxygen bridge in place and so inhibit reversibility; this explains why the addition of acid is required to recover the O_2 since this will destroy the hydroxo bridge and so presumably "unlock" the dioxygen. The formation of the μ-hydroxo bridge in basic media, resulting in the dibridged (O_2, OH) complex (*X*) therefore corresponds to an increase in the stability of the dioxygen adduct, but at the expense of preventing reversibility. These workers (*91*) also consider that the electronic structure of the dibridged product (*X*) is best characterized by a Co(III)-μ-hydroxo-μ-peroxo formulation, that is:

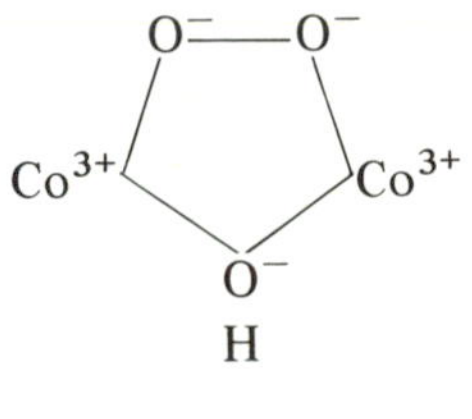

$$\text{Co}^{3+}\underset{\underset{\text{H}}{\text{O}^-}}{\overset{\text{O}^-\!-\!\text{O}^-}{\diamond}}\text{Co}^{3+}$$

A study (*92*) on the oxygenation of some cobalt(II)-polyamine chelates further clarifies the previous work on hydroxo bridge formation. It was found (*92*) that of the complexes studied only the diethylenetriamine cobalt(II) dioxygen complex was rapidly reversible toward inert atmosphere at both high and low pH, and toward EDTA addition at pH 9. It appears that the dissociation of the dibridged (O_2, OH) adducts is initiated by an attack at a labile (aquo) coordination site about the metal ion. Consequently, the oxygenated complexes that contain more than three nitrogen donors will not be readily reversible toward inert gases at low and high pH and EDTA addition. This explains the rapid reversibility of the diethylenetriamine cobalt(II) dioxygen complex. However, all the dibridged (O_2, OH) complexes could be dissociated by the addition of acid, via protonation of the hydroxo bridge. *Martell* and *Nakon* (*92*) also report that for the series of cobalt(II)-polyamine binuclear dioxygen complexes studied, the μ-hydroxo bridge is always present, except when the auxiliary ligands provide 5 or more strongly bonding (non-labile) donor groups per Co(II) atom, such as in the case of the singly bridged μ-peroxo binuclear tetraethylenepentamine cobalt(III) complex. That is, the presence of the hydroxo bridge appears to be dependent on steric factors; a point that has already been raised in section III (*B*) in respect to a study (*48*) of the binuclear oxygenated cobalt(II) complexes of some amines, amino acids and dipeptides. A recent study (*93*) of the effect of ligand basicity on the stability of some μ-peroxo-μ-hydroxo cobalt(II)-polyamine dioxygen complexes has shown that the logarithm of the stability constant of the dioxygen adduct bears a linear relationship to the sum of the pK values of the ligand(s) coordinated to the cobalt(II) ion. The investigation (*93*) reveals that both the symmetrical and unsymmetrical ethylenediaminediacetic acids form quite stable dibridged (O_2, OH) dioxygen complexes over an appropriate range of concentration and pH, even though they contain only two basic nitrogen atoms, rather than the three that have previously been claimed (*47, 48, 86*) as necessary. The linear free energy relationship given above demonstrates (*93*) that increased basicity of the ligands will lead to increased electron donation to the central metal, and this will, in turn, enhance the metal → dioxygen (π^*) back donation which appears to be necessary in the formation of a dioxygen adduct (for a discussion of the bonding interactions involved see section V(*B*)). Unfortunately, insufficient equilibrium data was available to enable a similar analysis of the effect of ligand basicity on the stability of monobridged μ-peroxo species, although some data was quoted (*93*). The cobalt(II) complexes of the two isomers N,N-bis(2-aminoethyl)-glycine (SDTMA) and diethylenetriamine-N-acetic acid (UDTMA) react (*84*) with molecular oxygen in aqueous solution to form binuclear dibridged (O_2, OH) species which can be deoxygenated by purging N_2 through a low pH solution, or by acidification. These dibridged (O_2, OH) dioxygen adducts are uncommonly stable towards oxidation or irreversible rearrangement to cobalt(III) complexes, and particular emphasis has been placed (*84*) on the complete reversibility of oxygenation on lowering either the O_2 concentration or the pH. The unusual stabilities may partly be due (*84*) to steric effects. The stability constant of the complexes formed by Co(II), Cu(II), Ni(II) and Zn(II) with SDTMA and UDTMA were determined potentiometrically, but interaction with dioxygen was only reported for the Co(II) complexes, as discussed above. A model peptide system Co(II)-DGENTA has been investigated, and reacts with molecular oxygen in aqueous media to form a monobridged μ-peroxo species (*94*). A decrease in either the O_2 pressure or the pH will lower the concentration of the oxygenated complex in solution, and purging with nitrogen results in completely reversible deoxygenation. An irreversible side reaction, probably to a Co(III) complex, was detected (*94*), but at such a slow rate that weeks would be required for appreciable reaction to occur at room temperature, compared with a few hours in the case of other cobalt-peptide complexes (*95*). The cobalt(II) complexes Co(terpyridine) $(LL)^{2+}$, where LL = 1,10-phenanthroline or 2,2′-bipyridine, are unique (*96*) in that they take up dioxygen reversibly in strongly acid solution, the uptake being complete at pH 3; a study of the equilibria involved during oxygenation, and an ESR analysis of the product, both support the conclusion that monobridged μ-superoxo complexes are formed. However these dioxygen adducts are not stable to reversible experiments (*96*).

A calorimetric study has been made (*97*) of the oxygenation of the cobalt(II) complexes formed with histidine, histamine and ethylenediamine in aqueous media at 25 °C.

The dioxygen adducts involving histidine and histamine contain a single dioxygen bridge, whereas in the case of ethylenediamine a dibridged (O_2, OH) product is formed, as already discussed. However, the oxygenation process to form either the monobridged or the dibridged adduct, is characterized by large negative enthalpy and entropy changes (*97*). The interaction of dioxygen with various cobalt(II) glycylglycine chelates has recently been investigated (*112*) and it appears that the difference in O_2 affinity found for various peptide derivatives may depend simply on electrostatic factors which could prohibit the formation of a monobridged μ-peroxo adduct, or in other cases (*113*) lead to irreversible oxygenation.

Much work has been carried out on *Werner* (*98*) complexes (see section III(*D*)). An ESR study (*99*) of the complex

$$[(CN)_5Co{-}O_2{-}Co(CN)_5]^{5-}, \qquad (C)$$

indicates the presence of a μ-peroxo bridge this is contrary to previous results (*100*). Recall that the paramagnetic complex

$$[(NH_3)_5{-}Co{-}O_2{-}Co(NH_3)_5]^{5+}, \qquad (B)$$

was also found (*68*) to contain a peroxo bridge, with the Co–O–O–Co moitey being planar (*69, 70*). The ESR spectrum (*102, 103*) of the dibridged complex

$$[(NH_3)_4Co{-}(O_2, NH_2){-}Co(NH_3)_4]^{4+}, \qquad (D)$$

suggests that the μ-peroxo bridge is best characterized as a modified superoxide ion. Also, equilibria studies (*101*) indicate that the μ-amido bridge in (*D*) contributes to a stabilization of the adduct, similar to that observed for μ-peroxo-μ-hydroxo complexes. An analogue of (*D*) containing ethylenediamine instead of ammonia has also been studied (*104, 105*). An X-ray analysis (*106*) of the diamagnetic complex

$$[(CN)_5Co{-}O_2{-}Co(CN)_5]^{6-}, \qquad (E)$$

shows that the Co–O–O–Co moitey has a staggered configuration and the O–O bond length is 0.1447 nm, this is typical of other peroxo compounds. A recent X-ray analysis of the paramagnetic complex (*C*) has also been completed (*107*). Two independent anions have been found (*107*) to exist in the (*C*) structure, in one case the Co–O–O–Co moitey is planar, and in the other case non-planar, however, in both anions the superoxo bridge is in a staggered configuration. The averaged O–O and Co–O bond lengths in (*C*) were found (*107*) to be 0.126 nm and 0.194 nm, respectively; the O–O bond length is characteristic of a superoxide grouping. By considering the structural data for the dioxygen complexes

$$[(NH_3)_5Co{-}O_2{-}Co(NH_3)_5]^{4+}, \quad \text{(refs. } 71, 72, 117\text{)}, \qquad (A)$$
$$[(NH_3)_5Co{-}O_2{-}Co(NH_3)_5]^{5+}, \quad \text{(refs. } 69, 70, 118\text{)}, \qquad (B)$$
$$[(CN)_5Co{-}O_2{-}Co(CN)_5]^{6-}, \quad \text{(ref. } 106\text{)}, \qquad (E)$$

and

$$[(CN)_5Co–O_2–Co(CN)_5]^{5-}, \qquad \text{(ref. } \mathit{107}\text{)}, \qquad (C)$$

certain generalizations have been made (*107*) regarding the preferred orientation of the dioxygen ligand:

1) peroxide groupings are non-planar when extensive hydrogen bonding is present and are planar otherwise,
2) superoxide groupings prefer a planar configuration, but will be non-planar if crystal packing requirements necessitate this, and
3) the ammine complexes have longer O–O and shorter Co–O bond lengths than in the corresponding (peroxo or superoxo) cyano complexes, since the π-acceptor ability of O_2, NH_3 and CN is in (*107*) the order $CN > O_2^- \approx O_2^{2-} > NH_3$;
this would result in Co–O > Co–N bonding strengths in (*A*) and (*B*) accompanied by strong back donation to the dioxygen π^* orbitals, thus explaining the lengthening of the O–O bond. Whereas we expect Co–C > Co–O bonding strengths in (*E*) and (*C*), which is similarly consistent (*107*) with the shorter O–O bond length found for these complexes. Analyses of the electronic spectra for the complexes (*B*) and (*C*) have been reported (*108–111*). The characteristic low energy bands at 672 nm for (*B*) and 486 nm for (*C*) have been assigned (*108*) to charge transfer transitions from the metal to the out of plane $\pi^*(O_2^-)$ orbital; the assignments being based on an MO diagram appropriate for a C_{2h} superoxo-dicobalt(III) complex in which the Co–O–O–Co moitey is planar.

Much work has been devoted to the study of Schiff base complexes, in particular M(salen), where M = metal, has been the subject of extensive work (*114*). The early work by *Calvin et al.* (section III(*A*)) suggested that the 2 : 1(M : O_2) dioxygen adduct, type I, formed by Co(salen) in the solid state, contains a peroxo linkage. An X-ray analysis (*115, 116*) of the complex $(Co \cdot Salen)_2O_2(DMF)_2$ supports this hypothesis; see Fig. 5 for the pertinent results of this study.

In this study the Co–O–O–Co grouping was found to be non-planar, which is suggestive of peroxo type bonding. However, the value of the O–O bond length, smaller than that expected for a pure peroxo linkage (0.148 nm) may be indicative of a partial transfer of electrons from cobalt to dioxygen (*115*); and this appears to be related, in general, to the reversibility of oxygenation. The dioxygen complex is thermally unstable at 100 °C (*122*) losing oxygen and solvent molecules. An inactive crystalline modification of Co(salen)

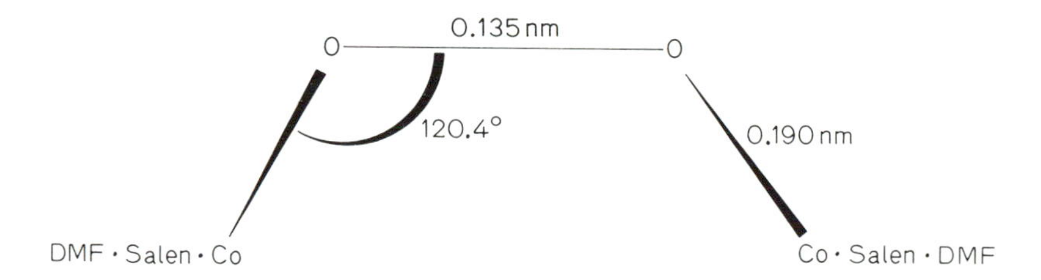

Fig. 5. $(Co \cdot Salen)_2O_2\,(DMF)_2$ (refs. 115, 116)

has been investigated by X-ray diffraction (*119, 120*), and found to exist of centrosymmetric dimeric units, $(Co\ salen)_2$, packed in such a way as to be self protecting towards O_2 uptake, rather than monomer units arranged in layers, as previously suggested (*17*); the dimers are bonded between the cobalt atoms and phenolic oxygen atoms of adjacent pairs of molecules. Also, an X-ray study (*121*) of Co(salen)Py attributes its inactivity in the crystalline state to too close packing.

We have already noted that binuclear dioxygen complexes are formed predominantly by cobalt(II) since, as already discussed, other metal(II) species, such as iron(II), will tend to be oxidized to metal(III) species. For example, the solid product recrystallized from an aerated pyridine solution of Fe(II) salen yields (*123*) the oxo-bridged complex $(salen \cdot Fe)_2O$; (2 pyridine), the structure of which has been obtained (*124*); the Fe–O bond length is 0.18 nm and the Fe–O–Fe bond angle = 139°. The iron is formally in the Fe(III) oxidation state in this dimer. Two forms of oxidation product have been isolated from aerated solutions of Mn(II) salen; Mn(III)(salen) OH (*125, 126*) and $(Mn(III)salen)_2O$, (H_2O) (ref. *126*). However, a more recent study (*127*) suggests the formation of three oxidation products containing $Mn(III)–O_2–Mn(III)$, $(Mn(IV)–O)_n$ and Mn(IV) = O, respectively; $(salen\ Mn(III))_2O_2^{2-}$ and $(Mn(IV)\ salen\ O)_n$ resulted from the oxygenation of solutions of Mn(II) salen in dimethyl sulphoxide and dimethylformamide or pyridine, respectively, and O=Mn(IV) (3-methoxy-salen), 1.5 CH_3OH from the oxygenation of a methanol solution of Mn(II) (3-methoxy-salen), H_2O. The monobridged μ-peroxo-manganese(III) adduct is formed irreversibly (*128*). An X-ray analysis (*129*) of the product isolated from an aerated pyridine solution of Mn(II) salpn shows the formation of a di-μ-hydroxo bridged adduct, and another manganese Schiff base complex has been prepared (*133*) which contains di-μ-oxo bridging. *Earnshaw et al.* (*131*) could find no relationship between the detailed magnetic properties and O_2 carrying ability of Co(II) salen, and a large number of its ring-substituted derivatives; including the 3-methoxy, 3-ethoxy and 5-nitro derivatives, which bind O_2 reversibly, and many inactive derivatives. An X-ray photoelectron spectroscopic study (*132*) of the binuclear cobalt(II) complexes

$$[Co(3–CH_3O–Salen)]_2O_2(DMSO)_2$$

and

$$[Co(5–Cl–Salen)]_2O_2(DMSO)_2$$

has been employed recently to estimate the binding energies of the cobalt core electrons in these complexes; the binding energies are found to increase for the oxygenated species, in respect to the unoxygenated species, and therefore indicates that the cobalt(II) becomes oxidized during oxygenation (oxidative addition), so that the above dioxygen adducts are best formulated as $Co(III)–O_2^{2-}–Co(III)$ species. But since these adducts remained (*132*) oxygenated at 10^{-6} torr pressure and room temperature (the experimental conditions for the study), it seems probable that the above electronic structure characterizes type I adducts of a kind not easily reversible with changes in pressure. The binding energies of the cobalt core electrons in $(Co \cdot Salen)_2O_2$, $(Co \cdot Salen)_2O_2(DMF)_2$ and $(Co \cdot 3\,MeO–Salen)_2$-$O_2(H_2O)_2$ were not determined in this study (*132*) because the adducts lost their O_2 under the vacuum conditions of the experiment.

B. Type II(P) Dioxygen Complexes

A far smaller number of type II(P) dioxygen adducts are reported in the literature as compared with type I adducts; we have already eluded to thermodynamic arguments (see beginning of previous section) which rationalize this disparity, at least in the case of cobalt(II) dioxygen adducts formed in aqueous solution, and we expect that iron(II) complexes will generally be oxidized to iron(III) species on oxygenation. In fact, in all known iron(II) complexes, reaction with dioxygen "in solution" is at least partly irreversible and leads (*135*), through autoxidation, to iron(III) species, whereas hemoglobin, which is known (*134*) to contain iron(II); exhibits the well known reversible behaviour with 1 : 1 stoichiometry, per iron atom. In view of this problem, *Baldwin* and *Huff* (*135*) considered the reaction of dioxygen with iron(II) species:

$$\mathrm{Fe(II)} + \mathrm{O_2} \underset{a}{\rightleftharpoons} \mathrm{Fe(II){-}O_2^{\bullet}} \xrightarrow[b]{\mathrm{Fe(II)}} \underset{\text{unstable}}{\mathrm{Fe(II){-}O{-}O{-}Fe(II)}}$$

$$\xrightarrow[b']{\text{autoxidation}} \mathrm{Fe(III)\ complex}$$

They felt that if the ligands associated with the Fe(II) were large enough, step b would be prevented and therefore reversible behaviour, via step a, could result. In cobalt(II) complexes, as we have seen, the termination step b does not necessarily lead to the autoxidation step b′, and indeed many reversible cobalt(II) type I dioxygen adducts have been isolated (see sections III and IV.A), but in iron(II) complexes it is apparent that b′ generally follows b (or the bridging species is formed irreversibly with a tendency to undergo autoxidation). *Baldwin* and *Huff* tested their hypothesis by synthesizing various iron(II) complexes of substituted dihydrooctaaza-(14)-annulenes; two pertinent examples (see Fig. 6) being those formed with 9,10-bridged-9,10-dihydroanthracenes and cyclohexane-1,2-dione, yielding (*a*) and (*b*), respectively. The reaction of these complexes with O_2 in solution was observed (*135*).

They found (*135*) that a toluene-pyridine (1% v/v) solution of (*a*) did take up dioxygen reversibly with 1 : 1 stoichiometry at – 85 °C; regeneration was demonstrated from the characteristic spectra for deoxygenated and oxygenated forms. No sign of the Fe(III) complex was seen, a separate complex having been prepared for spectral characterisation; although irreversible changes occurred at $T > -50$ °C. Complex (*b*), however, which contains a less bulky chelate, forms an irreversible type I (binuclear) dioxygen complex in a toluene-pyridine (4% v/v) solution at – 78 °C. This study shows the importance of steric factors in bimolecular termination steps (which lead to the formation of type I dioxygen adducts). An X-ray study of a cobalt : dioxygen (2 : 1) type I complex (*115*) indicates a Co–Co distance of 0.45 nm. A similar Fe–Fe distance would be available (*135*) at the closest approach of two molecules of complex (*b*) – hence (*b*) forms an irreversible type I dioxygen complex – but two molecules of complex (*a*) could not approach this close, and hence a reversible mononuclear dioxygen complex is formed. Autoxidation of the irreversible type I dioxygen adduct formed by (*b*) would probably occur if the temperature was allowed to rise, possibly leading to the formation of a μ-oxo dimer, Fe(III)–O–Fe(III), or

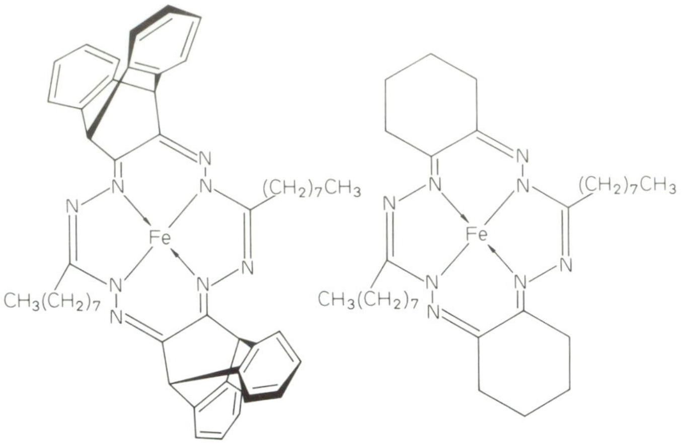

Fig. 6. Ferrous octaazamacrocyclic molecules.

some other Fe(III) species. In the case of vanadium(III) Salen there is interaction with molecular oxygen, but oxidation occurs, the oxo adduct VO(Salen) being the ultimate product (*163*).

We have seen, in the previous section (and section III.A), that cobalt (Salen) and its active derivatives normally form diamagnetic peroxo type I dioxygen adducts. However, a pyridine solution of Co(3-methoxy Salen) has been shown (*137*) to take up dioxygen with 1 : 1 (Co : O_2) stoichiometry; it was found (*137*) that there was no significant IR absorption band, attributable to the O–O stretch, for the oxygenated complex, and this suggested that the dioxygen is symmetrically bonded in an unidentate manner,

$$\begin{array}{c} O{=}O \\ \downarrow \\ Co \end{array}$$

since if there is no loss of symmetry about the dioxygen molecule, when it becomes co-ordinated, we would not expect it to be IR active. However, *Diemente et al.* (*138*) have carried out an ESR study confirming the formation of the 1 : 1 dioxygen adduct with Co(3-methoxy Salen), but which proposes the assignment of the unsymmetrical II(*P*) type dioxygen orientation. Further to this, IR (*139*) and ESR (*130*) studies of Co(acacen) (O_2) (base) both indicate that the dioxygen is unsymmetrically orientated, that is, a II(*P*) type adduct. *Diemente et al.* (*138*) have also found that Co(Salen) py forms a 1 : 1 dioxygen adduct both as a solid and in fluid and frozen solution. *Bailes* and *Calvin* (*11*) had originally shown that this complex forms a 2 : 1 dioxygen adduct, however, *Diemente et al.* (*138*) modified the original preparation by exposing to O_2 an anhydrous solution of Co(Salen) in pyridine at room temperature; the 2 : 1 dioxygen adduct (Co(Salen)py)$_2 \cdot O_2$ precipitated out over a period of several hours leaving a solution containing the 1 : 1 dioxygen adduct Co(Salen) (O_2) py; note that after storage for some weeks at room temperature the ESR signal characteristic of this 1 : 1 adduct disappeared. These workers (*138*) could find no evidence for an equilibrium between $(Co(Salen)py)_2 \cdot O_2$ and Co(Salen)(O_2)py. Other ESR studies (*140–142*) on oxygenated solutions of Co(Salen) also show the formation of a 1 : 1 dioxygen adduct, compatible with a II(*P*) orientation. The ESR signal produced was found to be dependent on the solvent used, and the evidence suggested (*141*) that the larger the crystal field strength (10Dq value) of the trans ligand (the solvent), the more stable will be the dioxygen adduct. The II(*P*) dioxygen adduct formed has been formulated (*141*) in a first approximation, as a $Co(III)–O_2^-$ complex. But in cases where the mononuclear dioxygen adducts of cobalt(II) Schiff base complexes undergo bimolecular termination in solution to form type I (binuclear) dioxygen adducts, an ESR study (*142*) has suggested the existence of an additional mononuclear dioxygen species, formed by the decomposition of the type I adduct; the complete scheme being given (*142*) by:

$$
\begin{aligned}
LCo(II) + O_2 &\rightleftharpoons LCo(III)–O_2^{\cdot} \\
LCo(III)–O_2^{\cdot} + LCo(II) &\rightleftharpoons LCo(III)–O–O–Co(III)L \\
LCo(III)–O–O–Co(III)L + O_2 &\rightleftharpoons 2\ LCo(II) \cdot O_2
\end{aligned}
$$

The $Co(III)–O_2^{\cdot}$ was assigned (*142*) to a narrow singlet ESR signal, whereas the $Co(II)O_2$ was assigned to an 8-line ESR signal, the latter being the ultimate product produced. However, a similar 8-line ESR signal produced (*130*) by an oxygenated solution of Co(acacen) has been assigned to the paramagnetic mononuclear dioxygen species containing the $Co(III)–O_2^{\cdot}$ moitey. The oxygenation process being thought (*130, 143*) to involve the transfer of an electron from the d_{z^2} orbital of the cobalt to one of the π_g orbitals of the oxygen molecule. The ESR data has also been discussed (*144*) in relation to the kinetics and thermodynamics of oxygenation of some cobalt Schiff base complexes, and this evidence supports the $Co(III)–O_2^-$ electronic structure for the mononuclear dioxygen adducts formed; the observed O–O stretching frequency of $\sim 1130\ cm^{-1}$, similar to that of O_2^-, is also consistent with this formulation (*151*). An ESR study (*145*) of the 1 : 1 dioxygen adduct formed by Co(II) tetrasulphopthalocyanine also favours the $Co(III)–O_2^-$ formulation. A minimal basis set ab initio molecular orbital calculation (*146*) on Co(acacen)L(O_2), where L = none, H_2O, imidazole, CN or CO, finds the $Co(III)–O_2^-$ electronic structure lowest in energy of four possible ground state configurations considered. An ESR spectral analysis (*147*) and an electrochemical study (*158*) of the dioxygen adducts formed by Co(II) Bae, Co(II) Salen and Co(II) Saloph in py/$CHC\ell_3$ solution confirms the formation of 1 : 1 (Co : O_2) adducts, which are best formulated (*147*) as $Co(III)–O_2^-$ species. This formulation is also favoured for the dioxygen complex Co(bzacen)py(O_2) for which an ESR spectrum has been obtained in both fluid (*148*) and frozen (*152*) solution, the O_2 having been ^{17}O enriched; the spin density on the dioxygen was found to be close to 100% as was also found in another ^{17}O ESR study (*149*) of dioxygen co-ordinated to a cobalt(II) – ammonia complex in γ-zeolite. An ESR study (*150*) of the complex Co(II)(Sal Me DPT), in $CH_2C\ell_2$ – toluene solution, and its reaction with O_2 and CO concludes that 1 : 1 adducts are formed in both cases with identical spectra, and moreover, the dioxygen is not formulated as superoxide ($Co(III)-O_2^-$), but as a singlet species. However, this system has been reinvestigated (*151*) and exposure to CO did not appear to lead to any observable interaction with CO or to an ESR spectrum resembling that for a mononuclear Cobalt-dioxygen adduct. This work (*151*) supports the assignment of a $Co(III)-O_2^-$ electronic structure for the mononuclear paramagnetic dioxygen adduct formed by Co(II) (Sal Me DPT) in solution as well as providing evidence for the formation of the diamagnetic μ-peroxo bridged adduct. The ^1H NMR of Co(II) (Sal Me DPT) on exposure to O_2 in $CDC\ell_3$ solution did not show a spectrum assignable to the mononuclear dioxygen adduct, although the slow appearance of a spectrum, assigned (*151*) to the diamagnetic μ-peroxo bridged species, was observed; this spectrum had previously (*150*) been attributed to the paramagnetic mononuclear dioxygen adduct.

It is clear from the ESR studies that have been reported that the majority of workers favour the $Co(III)-O_2^-$ formulation for most mononuclear type II(*P*) dioxygen adducts.

The X-ray analyses of a few of these adducts have been reported and confirm the bent structure for the Co–O–O grouping, and the O–O bond length is generally consistent with a superoxide species. ESR work (*130*) has already been presented which confirms that cobalt(acacen) forms a 1 : 1 dioxygen (superoxide) adduct in solution. Co(II)(acacen) (O_2) py has been isolated and an X-ray analysis (*153*) confirms that the Co–O–O grouping is bent, although the O–O bond length was not reported by these workers. An X-ray analysis (*154*) of an analogous dioxygen complex Co (bzacen)(O_2) py finds a Co–O–O angle of 126° and an O–O bond length of 0.126 nm, compared to 0.128 nm for pure superoxide (*1*), which supports the ESR work (*148, 152*) on this complex, discussed above, which itself concluded that the cobalt-dioxygen moitey exists as $Co(III)-O_2^-$, at least in a first approximation. The dioxygen adduct Co(Salen C_2H_4 py) (O_2) has been isolated (*155*) and X-ray analysis (*155, 156*) gives the Co–O–O angle and O–O bond length to be 136° and 0.11 nm, respectively. An X-ray analysis (cited in Ref. *156*) of the dioxygen adduct Co(t-Bsalten) (1-benzylimidazole) (O_2), finds the Co–O–O angle and O–O bond length to be 118.5° and 0.1251 nm, respectively; the O–O bond length agreeing well with that found in Co(bzacen) (O_2) py, again being consistent with a Cobalt(III)-superoxide formulation, however, the Co–O–O bond angle differs quite significantly in these two cases. Similar results have been obtained for the dioxygen adduct formed by "picket-fence" porphyrin, but we have postponed discussion of this until section V.

A further example is the dioxygen complex $[Co(CN)_5O_2]^{3-}$; X-ray analysis (*156, 157*) finds an O–O bond length of 0.124 nm which, although somewhat short, is consistent with a superoxide structure; the presence of an unusually large Co–O–O angle of 153° in this case can be rationalised (*156*) in terms of a strong repulsive interaction between the negatively charged terminal oxygen atom and four adjacent cyanide anions in $[Co(CN)_5O_2]^{3-}$, coupled with the weak Co–O–O angle bending force constant. Since the adjacent ligand atoms in neutral Schiff base dioxygen adducts, such as Co(bzacen) (py) (O_2) and Co(acacen) (py) (O_2), are either neutral or slightly positive there is no repulsion of their terminal oxygen atom, and therefore the Co–O–O angle is correspondingly smaller in these cases.

An X-ray photoelectron spectroscopic study (*132*) of some cobalt(II) Schiff base complexes and their dioxygen adducts finds that the cobalt core electron binding energies in the mononuclear dioxygen adducts investigated are consistent with the Co(III)–O_2^- formulation. A linear correlation has been found (*159, 160*) to exist between the tendency to form dioxygen adducts and the ease of oxidation of cobalt(II) to cobalt(III), as measured by cyclic voltammetry, for a series of complexes Co(II)LB, where L = tetradentate Schiff base and *B* = axial base. It was suggested (*160*) that this correlation exists because the redox potential of the cobalt chelate provides a measure of the electron density on the cobalt, which appears to be the major factor in determining O_2 affinity; increased electron donor properties of either the axial base *B* or the equitorial tetradentate ligand L resulted in a corresponding increase in O_2 uptake ability. Hence, the lower O_2 affinity exhibited (*161*) by Co(II)Salen in comparison to Co(II) Bae results from the electron withdrawing effect of the aromatic rings in the former complex, and this emphasizes that the electron donating (or withdrawing) ability of the equitorial ligands must be considered in addition to that of the axial base, which has generally received more attention. But an ESR study (*162*) of various reversible Schiff base type mononuclear dioxygen adducts shows that ESR parameters are insensitive to changes in axial ligands, which is in contrast to the 'sensitivity' found in an ESR study (*58*) of cobalt(II) corrinoids. Co(II)Bae was oxygenated (*59*) in a toluene solution containing pyridine or substituted pyridine at – 10 °C; as the base strength of the substituted pyridine increased, so did the formation tendency of the dioxygen adduct. *Collman et al.* (*155*) suggest that the rarity of complexes containing a metal surrounded by a pentaco-ordinate environment consisting of a planar tetradentate grouping and an axial base, may account for the relative scarcity of mononuclear dioxygen, type II(*P*), complexes.

C. Type II(G) Dioxygen Complexes

An illustration of the II(*G*) orientation for the dioxygen ligand is given in Fig. 2. We have not included any work on type II(*G*) dioxygen complexes in section III even though some of this work is over ten years old. The reason for this is that II(*G*) dioxygen complexes tend to be amenable to X-ray analysis and other physical investigations, leading to a definitive characterisation of these complexes, and are therefore considered in this section. Also, we have seen that type I and II(*P*) dioxygen complexes are often formed together in solution; but there is no case reported to our knowledge of any interconversion between II(*G*) dioxygen complexes with either I or II(*P*) dioxygen complexes. Consequently, this section is largely self-contained; except for the discussion of the bonding interactions involved which are discussed in section V. B. A large number of irreversible type II(*G*) dioxygen complexes are reported in the literature, however, we only include those examples needed to illustrate the principles appertaining to reversible oxygenation for type II(*G*) dioxygen adducts; a full catalogue of these irreversible dioxygen adducts is given elsewhere (*1, 77, 80*) in addition to an account of the reactions they undergo.

In 1963 *Vaska* (*164*) discovered that the iridium complex $Ir(PPh_3)_2C\ell(CO)$ takes up molecular oxygen reversibly with 1 : 1 stoichiometry. This complex has since been shown to reversibly sorb (1 : 1) ethylene (*165*), carbon dioxide (*166*), $F_2C{=}CF_2$ and $F_3C{-}C{\equiv}C{-}CF_3$ (*167*), as well as various other ligands (*168*). *Ibers* and *La Placa* (*169*)

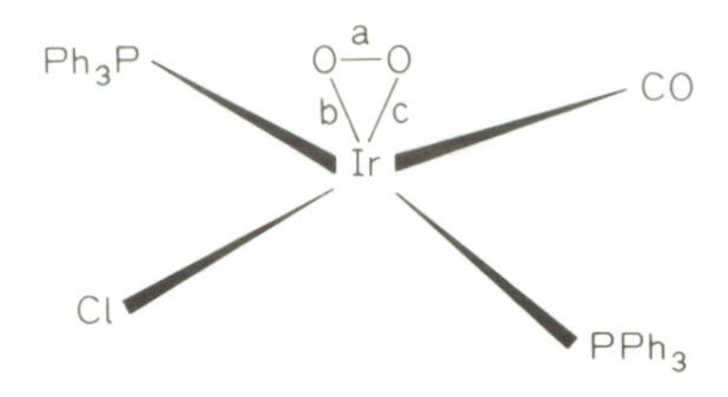

Fig. 7. Dioxygen adduct of Vaska's complex (Ref. 169).
(O–O), a = 0.130 nm
(Ir–O), b = 0.204 nm
c = 0.209 nm

have carried out an X-ray analysis of $Ir(PPh_3)_2(O_2)C\ell(CO)$; the pertinent results of this study are summarised in Fig. 7.

Ibers et al. (*170*) have determined the structure of the irreversible dioxygen complex $[Ir(PPh_3)_2(O_2)I(CO)]$ by X-ray crystallography and have found that the O–O bond length is 0.151 nm. These results were explained (*170*) as follows: As the halide becomes more electron releasing the greater is the back-donation from the metal to the O_2 antibonding molecular orbitals and consequently the longer the O–O distance. Therefore it was suggested (*170*) that the replacement of the chlorine in (VI) by iodine is responsible for the increase in the O–O bond length from 0.130 nm to 0.151 nm, which is roughly equivalent to a change from superoxide to peroxide bonding (see Table 1). It is now clear (*179*), that disorder in the crystals may be of significant influence in these results, as will be discussed later, and the conclusions of Table 1 are in doubt.

Table 1

Compound	O–O bond length nm	Ref.
O_2^+ (bond order 2.5)	0.112	(*77*)
O_2 (bond order 2)	0.121	(*77*)
O_2^- (bond order 1.5) superoxide	0.128	(*77*)
O_2^{2-} (bond order 1) peroxide	0.149	(*77*)
$Ir(PPh_3)_2(O_2)C\ell(CO)$ abbrev. (V1)	0.130	(*169*)
$Ir(PPh_3)_2(O_2)I(CO)$ abbrev. (V2)	0.151	(*170*)

Since (V1) is reversible and (V2) is irreversible, it was suggested (*170*) that a correlation could be made between the O–O bond length in dioxygen complexes and M–O bond strength with the degree of reversibility of their formation reactions (*1, 171*). *Vaska et al.* (*172*) have determined the structure of the diamagnetic irreversible II(*G*) dioxygen adduct $(O_2)Co(2 = Phos)_2BF_4$, and found an O–O bond length of 0.142 nm – this seems to be in agreement with the above hypothesis. However, the O–O bond length in the reversible dioxygen adduct

$$[(PPh_3Et)_2Ir(O_2)C\ell(CO)]$$

was found (*173*) to be 0.146 nm, which is about the same as in the irreversible dioxygen complex (V2). So it appears that there is no simple one-to-one correlation between reversibility and O–O bond length as previously suggested (*170*). We shall return to this point later. The kinetics of formation for a series of Vaska type dioxygen adducts have been investigated (*174–176*) and the rates of oxygenation and deoxygenation have been found to increase and decrease, respectively, with increasing basicity of the anionic ligand (A) in the series of complexes $Ir(A)(Ph_3P)_2(CO)$, provided the ligands (A) have comparable structures. The value of k_2 (the second order rate constant) for the complexes with (A) = Cℓ, Br and I was found to be in the order Cℓ < Br < I. We can correlate the basicity of the ligand (A), in general, with the CO stretching frequency (*168*), since electron transfer from (A) to Ir will tend to decrease the multiple bond character of the CO bond, through Ir → CO back-donation, and hence decrease the stretching frequency. It was found that the geometry of the ligand (A) exerts a profound effect on the dynamics of oxygenation (*174*), in that steric factors were sometimes observed to over-ride electronic considerations. *Ibers et al.* obtained (*177*) the structural parameters summarised in Table 2.

Table 2 (Ref. *177*)

Compound	O–O(nm)	M–O(nm)	O_2 uptake
$[Rh(O_2)(diphos)_2][PF_6]$	0.1418	0.2026 0.2025	reversible
$[Ir(O_2)(diphos)_2][PF_6]$	0.1625	0.1961 0.1990	irreversible

This data, taken in conjunction with previous work, led these workers (*177*) to the conclusion that increased electron density at the metal, either by substitution of a better donating ligand (P > I > Cℓ), or by changing the metal (Ir > Rh > Co), will ensure increasing uptake properties. This is consistent with an investigation (*178*) of $(PPh_3)_2Pt(O_2)$, using electron spectroscopy, which has shown that irreversible uptake of dioxygen is accompanied by a substantial electron transfer from metal to dioxygen, which is presumably enhanced by increased electron density at the metal; the O–O bond length in this complex being found to be (*183*) 0.145 nm.

The structure of $[Ir(O_2)(diphos)_2][PF_6]$ has been redetermined (*179*) and the results are found to be very different from those obtained (*177*) earlier. The M–O bonds, which are chemically equivalent, are now (*179*) found to be 0.231 and 0.234 nm, respectively, and are not substantially different, contrary to the earlier results shown in Table 2. Also, the O–O bond length previously found (*177*) to be 0.1625 nm, which is long, is now given (*179*) as 0.152 nm. Moreover, it has been concluded (*179*) from this and other work that the O–O bond length is essentially independent of the metal and ligand involved, usually being found in the approximate range 0.141 to 0.152 nm. If these assertions prove valid, we shall need to question more closely comparisons made between the O–O bond lengths for the Ir and Rh complexes in table 2, and between other dioxygen adducts, especially in view of the observed disorder in the crystals which is believed (*179*) to be responsible for the low value of 0.130 nm reported (*169*) for the O–O bond length in (V1), lying outside the 0.141–0.152 nm range expected (*179*).

Further to this, *Vaska et al.* (*181*) have studied the kinetics of dioxygen uptake for a series of d^8 isostructural, square planar (*182*), complexes M(2 = phos)$_2$(A); where M = Co, Rh, Ir and (A) = Anionic ligand. The observed order of attraction towards molecular oxygen was Co $\gg$ Ir $>$ Rh, which is not in agreement with the hypothesis discussed earlier which states that increased electron density at the metal should lead to increased uptake properties. These workers (*181*) were able to rationalise this unexpected order by suggesting that the activation enthalpies for oxygenation (Co $<$ Ir $<$ Rh) are proportional to the energy of an electronic transition in the reacting complexes, and if these energies are related to the $d(xy) \rightarrow d(x^2-y^2)$ transition, then it was suggested (*181*) that the activity of a particular metal towards dioxygen could be predicted in terms of its ligand field stabilization energy, which in turn could be correlated with electronic spectra.

Infra-red absorption bands have been reported for a wide range of dioxygen complexes (*77*). In general, the only band observed for type II(G) complexes is in the range 800–900 cm^{-1}, which has been assigned to the O–O stretch (*164*). This assignment has been confirmed for

$$[Ir(PPh_3)_2(O_2)C\ell(CO)] \text{ (Ref. } \mathit{184})$$

and for M(O_2) (t–BuNC) (Refs. *185, 186*), where M = Ni, Pd, by isotopic studies. Attempts have been made to correlate the O–O stretching frequency with the reversibility of dioxygen uptake.

Table 3 (Refs. *1, 77*)

Compound	O–O stretching frequency (cm^{-1})	O–O bond length (nm)
$Ir(Ph_3P)_2(O_2)C\ell(CO)$	858	0.130
$Ir(Ph_3P)_2(O_2)Br(CO)$	862	–
$Ir(Ph_3P)_2(O_2)I(CO)$	862	0.151

It can be seen from Table 3 that the change in O–O bond length, when Cℓ is replaced by I, is not accompanied by a corresponding change in the O–O stretching frequency. *Otsuka* (*185*), after calculating the isotopic splitting and force constants for the simple harmonic oscillators $^{18}O_2$, $^{18}O\,^{16}O$ and $^{16}O_2$, concluded from the observed spectra that the band at 800–900 cm^{-1} can not be solely due to a 'pure' O–O stretch. Other workers (*187, 188*) have arrived at similar conclusions, and it seems well established that a symmetric metal-oxygen stretch has an effect on the O–O stretching frequency, implying substantial M–O bonding. The experimentally observed constancy of the O–O stretching frequency is a consequence of two opposing factors (*77, 185*): An increase in the O–O force constant will invariably lead to a corresponding reduction in the M–O force constant, and consequently the O–O stretching frequency will not vary significantly, and is therefore a poor guide to reversibility or O–O bond length.

Huber et al. (*189*) have investigated cocondensation (4.2–10 °K) reactions between Ni, Pd and Pt and molecular oxygen in pure O_2 and O_2/Ar matrices. These reactions were studied by matrix isolation infra-red spectroscopy, including isotopic and diffusion controlled warm-up studies. They established that both $M(O_2)$ and $(O_2)M(O_2)$ species were present. The O–O bond order suggested significant back-bonding, and this led them to reject the monodentate structure, Fig. 8 (b).

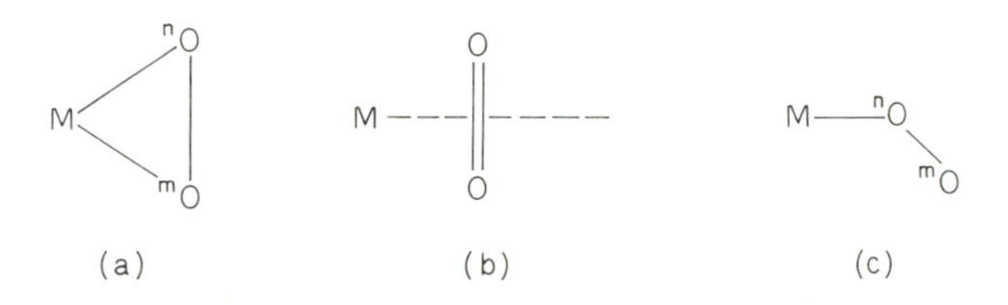

Fig. 8. n, m = 16 or 18 (Ref. *189*)

They found that the order for the M–O stretching modes (M–O force constants) was $(O_2)Pd(O_2) > Pd(O_2)$ and $(O_2)Ni(O_2) > Ni(O_2)$, and since the O–O force constant increased as the M–O force constant decreased, reversibility could not be equated with shorter O–O bond length. To test the proposed isosceles model they predicted the various absorptions that would be expected on the basis of the symmetrical structure (a) and unsymmetrical structure (c), Fig. 8. If the isotopic ratio of $^{16}O : {}^{18}O$ is 1 : 1, then three bands would be expected for structure (a), with relative intensities 1 : 2 : 1, and four bands would be expected for structure (c), with relative intensities 1 : 1 : 1 : 1, the two oxygen atoms now being in different environments. These workers (*189*) then obtained the IR spectrum for the cocondensation product from the reaction between nickel and O_2 (4.2–10 °K), in isotopic ratio $^{18}O : {}^{16}O = 1 : 1$, and found three bands with relative intensities 1 : 2 : 1, as predicted on the basis of the isosceles model Fig. 8 (a), that is, the II(*G*) orientation Fig. 2. This work has been extended recently (*190*) through a similar investigation of the binary mixed dioxygen dinitrogen complexes of nickel (*191*), palladium and platinum, $(O_2)M(N_2)_n$, where n = 1 or 2, formed from the cocondensation reactions of Ni, Pd and Pt atoms with mixtures of O_2, N_2 and Ar at 6–10 °K. The dioxygen ligands are again bonded side-on in the II(*G*) orientation, but the dinitrogen ligands are bonded end-on in a linear manner.

Bayer and *Schretzmann* (*25*) came to the conclusion that reversible oxygenation is a characteristic property of group VIII metals. However, work has shown that the cadmium complex $CdEt_2$ can take up dioxygen reversibly in the ratio 1 : 2 ($Cd : O_2$). But it was found that the oxygenated complex (II(*P*) or II(*G*) orientation) can undergo spontaneous catalytic oxidation to form bis(ethylperoxy) cadmium:

$$2\,O_2 + CdEt_2 \rightleftharpoons CdEt_2 \cdot 2\,O_2 \rightarrow (EtOO)_2Cd$$

A recent and novel example of a II(*G*) dioxygen adduct is that formed (*193*) by $Co(CN)_2(PMe_2Ph)_3$, thus:

$$2\,Co(CN)_2(PMe_2Ph)_3 + O_2 \xrightarrow{\text{benzene solution}} Co_2(CN)_4(PMe_2Ph)_5(O_2) + PMe_2Ph$$

At first glance this dioxygen complex may have been thought to be an O_2 – bridged type I adduct, however, an X-ray analysis (*193*) shows that the cobalt atoms are bridged through a CN^- ion, and the dioxygen is bonded in a II(*G*) orientation, to a single cobalt atom, with an O–O bond length of 0.144 nm.

Finally, it should be noted that there are a large number of irreversible II(*G*) dioxygen complexes (see Refs. *1, 77*) which undergo a whole range of oxidative reactions with a variety of substrates. For instance, $M(Ph_3P)_4$, where M = Ni, Pd, Pt, react with O_2 in solution as follows (*194*).

$$M(Ph_3P)_4 \rightleftharpoons M(Ph_3P)_2 + 2\,Ph_3P$$
$$M(Ph_3P)_2 \xrightarrow{O_2} (Ph_3P)_2M(O_2)$$

The dioxygen adduct so formed is highly reactive and will catalyse the oxidation of phosphine to phosphine oxide and isocyanide to isocyanate (*77*). But clearly a survey of these reactions is not within the scope of this review and is provided elsewhere (*1, 77*).

V. Hemoglobin and Synthetic Metal Porphyrins.

A. Introduction

The most important of all known oxygen-carriers are the naturally occurring molecules hemoglobin and myoglobin. Hemoglobin has a molecular weight of approximately 64,000 and consists of four subunits, each of which contains one haem group (*195*). The structures of hemoglobin and myoglobin have been determined by *Perutz* (*196*) and *Kendrew* (*197*), respectively, using X-ray crystallography. Myoglobin contains one haem group and is very similar in structure to a subunit of hemoglobin. The iron in hemoglobin is generally ferrous (although ferric forms are known), the complex being paramagnetic (*134*); oxyhemoglobin, however, is diamagnetic. This is of great interest, since it involves important implications for the structure of HbO_2 (*32*). The haem group is attached to the protein in both hemoglobin and myoglobin through a co-ordinated histidine-nitrogen atom (*195*), and trans to this position is a vacant site which is occupied by a dioxygen ligand in the oxygenated species. Thus, the ligand-field in HbO_2 is approximately octahedral, and hence the para-

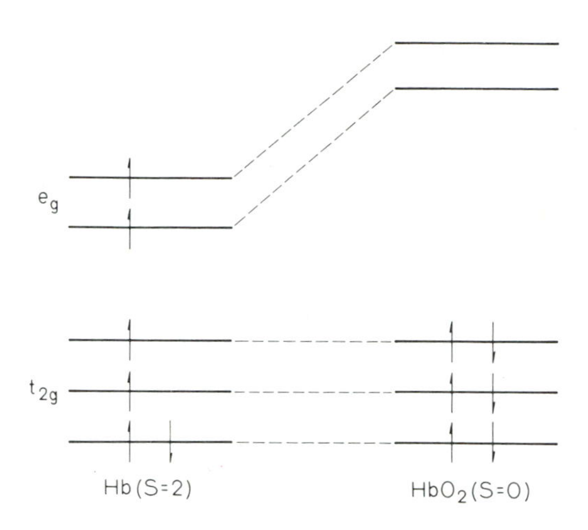

Fig. 9. Crystal-Field Energy level diagram for Fe(II)

magnetism of HbO_2 can be explained, for simplicity, in terms of the energy level diagram given in Fig. 9.

In Hb there are two electrons in the $\mathbf{e}_g$ set, and consequently the bonding radius of the iron is too large to fit into the plane of the nitrogen atoms of the haem porphyrin. The iron atom is found (*198*) to lie about 0.08 nm below the porphyrin ring in Hb. When the dioxygen is bound, the $\mathbf{e}_g$ orbitals are directed towards the ligands, which now make up an octahedral environment, and this therefore increases their energy considerably, hence the 3*d*-electrons in HbO_2 pair up in the $\mathbf{t}_{2g}$ orbital set, and consequently HbO_2 is diamagnetic. Since the $\mathbf{e}_g$ orbitals are now empty, the radius of the iron decreases so that it now fits into the plane of the porphyrin (*196*). A similar movement of the cobalt atom in coboglobin, the cobalt analogue of hemoglobin, occurs during oxygenation; the cobalt lies 0.038 nm out-of-plane in coboglobin and lies in plane in oxycoboglobin, the movement being about half that found for hemoglobin (*199*). The stereochemistry of co-operative effects in both hemoglobin (*198, 201*) and coboglobin (*199*) have been investigated. It is a well known but interesting phenomenon that the more O_2 taken up by Hb the greater is its affinity towards further O_2 uptake. So that if an O_2 molecule were to approach two hemoglobin molecules, one with three dioxygen ligands already co-ordinated, Hb‴, and the other with no O_2 bound, Hb, the chances are 70 to 1 in favour of the free O_2 molecule becoming co-ordinated to the Hb‴ molecule in preference to the Hb molecule (*198*). This effect is attributed to haem-haem interaction (*195, 198*). Hemoglobin is constrained by salt-bridges which are believed (*198*) to be broken by the energy of haem-haem interaction, with the release of hydrogen ions. The Bohr effect (*195*) results (*200*) from changes in the quaternary structure which alters the environment of three pairs of weak bases so that they tend to take up hydrogen ions on release of O_2 from oxyhemoglobin resulting in corresponding changes in pH; myoglobin shows no Bohr effect (*226*). Studies of the changes in concentration and binding of 2,3-diphosphoglycerate in hemoglobin during oxygenation, is also of considerable interest (*198, 201*). However, the study of the various effects which accompany the oxygenation process in hemoglobin presents a formidable volume of research in itself (see, for example, Refs. *25, 95, 195, 200, 201*), and will therefore not be discussed further in this review, except for changes occurring directly at the haem group. The rather more ambitious objective of synthesizing a model porphyrin which can reversibly bind O_2 under normal conditions, and which forms a dioxygen adduct that can be isolated for the purpose of X-ray analysis, has only recently achieved any significant success (see section V.C). It was reported (*203*) a long time ago that imidazole ferroprotoporphyrin takes up O_2 reversibly in the crystalline state; an attempt to prepare the complex in aqueous solution was unsuccessful. Also, *Wang* (*204*) reported spectral evidence that O_2 binds reversibly to 1-(2-phenylethyl) imidazoleheme diethyl ester embedded in a matrix of polystyrene and and 1-(2-phenylethyl) imidazole, but definitive structural evidence was not obtained. More recently (*233*) it has been reported that cross-linked polystyrene containing imidazole ligands did not provide a support rigid enough to prevent dimerization, and that the μ-oxo dimer was benzene extracted from oxygenated tetraphenyl porphyrin iron(II), Fe(TPP), which had been attached to the modified polystyrene. A discussion of model synthetic porphyrins, from which definitive structural and other physical data are obtained, is given in section V.C.

B. Theoretical Considerations Concerning the Orientation and Electronic Structure of the Dioxygen Ligand.

With respect to the orientation of the dioxygen ligand, *Pauling* and *Coryell* (*134*) originally proposed a linear structure for Fe–O–O in oxyhemoglobin. However, the spin-only magnetic moment indicates that the symmetry around the O_2axis in the complex has been sufficiently reduced to remove the degeneracy of the dioxygen antibonding orbitals; the Fe–O–O linear orientation is therefore untenable (*30, 32*). *Griffith* (*32*) did not specify in advance the orientation of the O_2 internuclear axis relative to the Hb, but introduced a coordinate system Q_{xyz}, as shown in Fig. 10, to describe the O_2 valence state. We can consider each oxygen atom (by analogy with ethylene) to be sp^2 hybridized. And it follows (*32*) that there are two sets of directions in which electron density is a maximum.

The first region of maximum electron density is along the lone-pair directions, and the second is in the Q_{yz} plane, parallel to the Q_z axis. Thus, two possible modes of coordination arise, these are shown in Fig. 11, which illustrates the proposed bonding in HbO_2.

Fig. 11(c) shows the sigma-type interaction between a filled O_2 π-orbital and the empty d^2sp^3 hybrid orbital on the iron atom. And Fig. 11(b) shows the π-type interaction between a filled *d*-orbital on the metal (*e.g.* d_{yz}) and one of the empty π^* dioxygen orbitals. To estimate the accessibility of electrons in the dioxygen molecule for structures II(*P*) and II(*G*), *Griffith* (*32*) considered the ionization energies of the donating orbitals involved in each structure. In the structure II(*P*) we therefore need to consider the ionization energy of a one-pair sp^2 hybrid orbital; a particular one of these is (*32*):

$$\chi = \frac{1}{\sqrt{3}}\Psi_{2S} - \frac{1}{\sqrt{2}}\Psi_{2P_x} + \frac{1}{\sqrt{6}}\Psi_{2P_y}$$

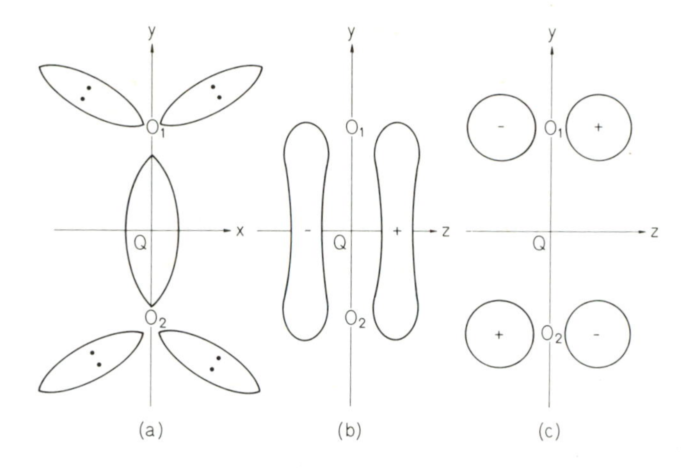

Fig. 10.
(a) The four lone-pairs and the sigma bond orbital
(b) Occupied bonding π-orbital
(c) Empty anti-bonding π^*-orbital (Ref. *32*)

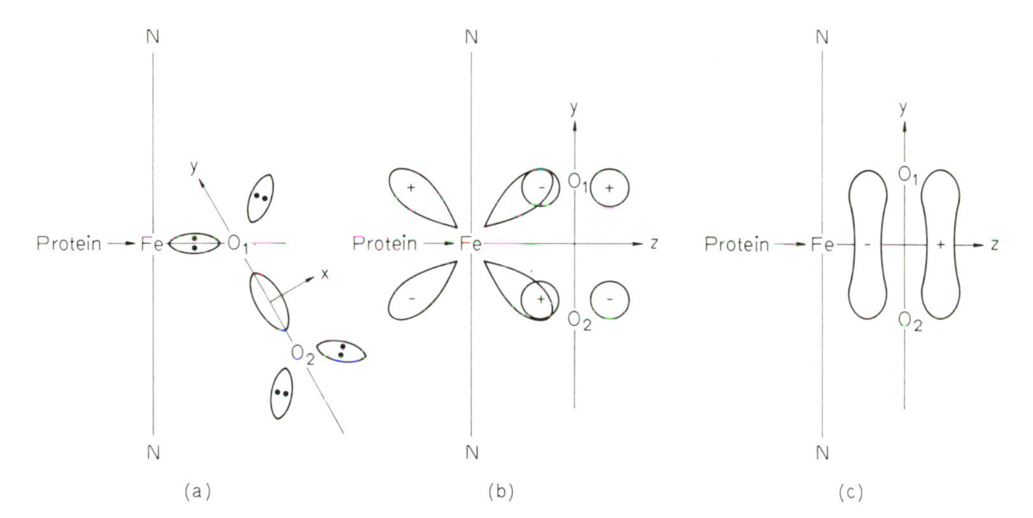

Fig. 11.
(a) The II(P) orientation,
(b) and (c) the II(G) orientation (Ref. *32*)

this orbital is located entirely on one of the oxygen atoms. We can therefore estimate the ionization energy of χ to have one-third $2\,s$ and two-thirds $2p$ character, hence

$$I(\chi) = \tfrac{1}{3} \cdot I(2\,s) + \tfrac{2}{3} \cdot I(2p)\,.$$

By using atomic spectral data to estimate the values of I(2 *s*) and I(2 *p*), *Griffith* (*32*) found I(χ) = 0.216 kJ/Mole. With structure II(*G*) we have to employ molecular spectral data. Applying similar approximations as were used in the previous case, *Griffith* (*32*) found I(π) = 0.172 kJ/mole. Now since this calculation gives I(χ) to be 0.044 kJ/mole less than I(χ), it was concluded (*32*) that structure II(*G*) is significantly more favourable than structure II(*P*), even though simplifying assumptions were made. Griffith's model is perhaps best illustrated by Fig. 12. In the case of the ligand NO^-, which is isoelectronic with O_2, coordination to both iridium (*208, 210*) and cobalt (*209*) occurs in a II(*P*) type orientation, bonded through the nitrogen, since for NO^- the ionization energy of a nitrogen lone-pair electron is smaller than that of a π-electron (*210*); the NO^+ ligand, however, coordinates in an approximately linear manner (*211*).

Three situations can be considered (*1*) for the bonding of dioxygen to the metal on the basis of Fig. 12. Firstly, when the metal orbitals lie as high or higher in energy than the π^*-orbitals in dioxygen (case *A*, Fig. 13), the bonding orbitals formed from the metal orbitals and dioxygen π-orbitals will be mostly dioxygen in character. Therefore there will be little dioxygen to metal sigma-type bonding. Also, the bonding orbitals produced between the metal and oxygen atoms will have both metal and oxygen character and will therefore produce strong M–O bonding. And if the back-donation is large enough the O–O bond will effectively become a single bond.

Conversely, the relative energies of the orbitals in case *C* will favour $\pi \rightarrow$ metal electron donation leading to largely sigma-type bonding from the dioxygen to the metal with little lengthening of the O–O bond. Griffith's theory is often illustrated by case *C*, but this neglects the essential element of back-bonding, and so misrepresents his theory. The intermediate sketch, case *B*, probably represents the situation best for a reversibly formed II(*G*) dioxygen adduct if a stick-diagram has to be used. We can consider *A* and *C* as limiting cases, *A* probably representing an irreversible II(*G*) dioxygen adduct. It has been noted

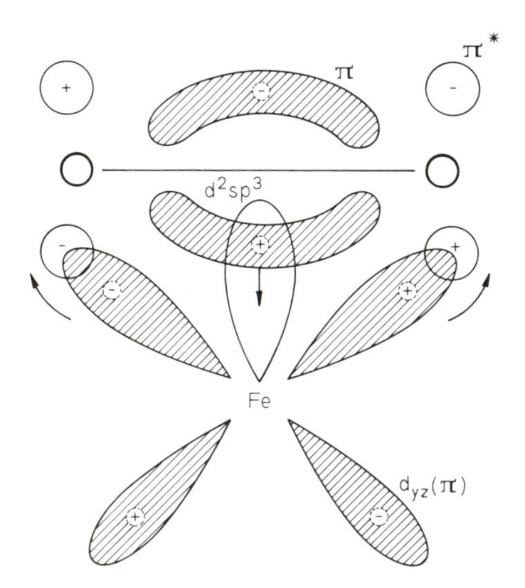

Fig. 12. Griffith's model; shows $(O_2)\pi \rightarrow$ metal ($d^2\,sp^3$ hybrid) donation, and $d(\pi) \rightarrow (O_2)\,\pi^*$ back-donation (Ref. *1*)

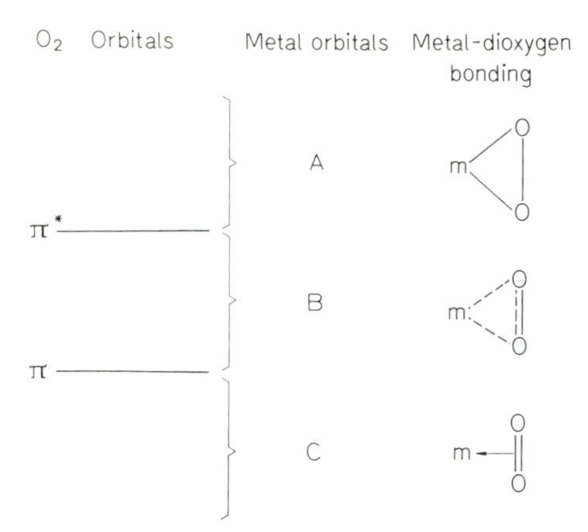

Fig. 13. Relative energies of O_2 and Metal Orbitals (Ref. *1*)

(*205*) that when a transition metal is in a relatively high oxidation state, the *d*-orbitals will be considerably contracted compared with the zero-valent state; overlap with antibonding π^* O_2 orbitals (as in Fig. 11(b)) will therefore be small and thus the II(*P*) arrangement favoured. Therefore, the dioxygen ligand coordinated to iron(II) and cobalt(II) complexes seems to favour the II(*P*) orientation, as opposed to the II(*G*) orientation, because the filled metal *d*-orbitals are contracted too much to allow sufficient overlap with the π^* dioxygen orbitals. In iridium, rhodium and platinum complexes, which coordinate dioxygen in the II(*G*) orientation, the filled back-bonding metal orbitals are presumably not contracted to this extent, being in relatively low oxidation states. *Mingos* (*206*) has used theory based on second order Jahn-Teller distortions to rationalize the preference for II(*G*) dioxygen bonding in rhodium, iridium and platinum complexes. The preferred orientation being thought (*207*) to depend solely on the direct product of the symmetries of the donor and acceptor orbitals involved. It is stated (*207*) that if the donor and acceptor orbitals have the same symmetry their direct product will be totally symmetric and no distortions which will change the point group of the complex are allowed; but if they have different symmetries the direct product will have the same symmetry as a normal vibrational mode of the complex which would ultimately result in a new orientation. These symmetry arguments are used (*207*) to support the II(*G*) assignment for the dioxygen orientation in HbO_2, but it is also stressed that nonbonded repulsions (*202*) between the π-dioxygen and porphyrin nitrogen atoms may also be important in determining the preferred orientation. We have seen that mononuclear dioxygen adducts of iridium complexes, for instance, are invariably type II(*G*), whereas those of cobalt complexes are usually type II(*P*). In view of this it is suggested (*212*) that the iridium valence orbitals in these complexes are probably at an energy, relative to the dioxygen orbitals, characterized by either case *A* or *B* in Fig. 13. Whereas in the cobalt complexes the valence orbitals of the metal are probably at a relative energy characterized by case *C*, so that the unidentate perpendicular structure, Fig. 13, would be predicted. However, the metal → dioxygen back-donation is probably essential (*212*) in order to counteract the polarity of the metal-dioxygen moitey produced by the initial dioxygen → metal electron transfer, and also the dioxygen adduct is unlikely to be stable if there does not exist at least some bonding interaction along or about the M–O internuclear axis (or axes). Now since in this instance, case *C* Fig. 13, the metal → dioxygen back donation is energetically prohibited, the only possible alternative bonding mode available to the dioxygen, leading to a stable $M-O_2$ moitey, is the II(*P*) orientation, and it is therefore energetically favoured in this case. Given the appropriate assumptions regarding the energy of the metal valence orbitals, relative to the dioxygen orbitals, these qualitative arguments appear to explain why the mononuclear dioxygen complexes formed by cobalt and iron are usually found to be type II(*P*) adducts, whereas those of iridium, rhodium and platinum are invariably type II(*G*) adducts – though it would be wrong to attempt to classify all dioxygen complexes of third row metals as type II(*P*) adducts since many nickel complexes are found (*77*) to form type II(*G*) dioxygen adducts. The above criterion, based on the relative energies of the dioxygen and metal valence orbitals, can be used in conjunction with the criterion discussed earlier, which was based on the extent of contraction of the metal back-bonding orbitals, to predict the preferred orientation for a particular dioxygen adduct. Hopefully a theory can be developed which not only quantifies the above arguments, but also enables a calculation of the perturbations produced by a particular ligand system to be made.

Weiss (*213*) proposed that the dioxygen in oxyhemoglobin is in the form of a superoxide ion, with the iron being in a ferric state, i.e. $Fe(III)-O_2^-$. He based this on the hypothesis that there was a large iron to dioxygen electron transfer. *Viale et al.* (*214*) carried out MO calculations which imply that the dioxygen ligand donates an electron pair to the iron, which then back donates; this back-donation being large. However, these calculations were based on the II(*P*) structure for the dioxygen ligand, no other structure was considered. We should not, therefore, adopt the II(*P*) structure in preference to the II(*G*) structure, simply on the basis of these calculations. *Pauling*, who first proposed the bent II(*P*) structure (*30*), does not agree with Weiss, and suggests (*31*) a charge separation within the dioxygen ligand itself i.e. $Fe(II)-O^+-O^-$(bent).

Halton (*215*) has carried out SCC–EH–MO calculations on the bonding and EFG in oxyhemoglobin. She considers several geometries, including II(*G*) and II(*P*), and finds that the largest 'net overlap population' for the dioxygen ligand is predicted for the II(*P*) orientation, which is therefore considered as the most favourable structure for $Fe(O_2)$ in oxyhemoglobin.

However, although she does deduce (*215*) a significant O–O charge separation, it is much smaller than that originally suggested by *Pauling* (*31*). *Halton* could only achieve satisfactory convergence in these calculations (*215*) if she assumed that there were two extra electrons in the iron-dioxygen system. It was presumed that these electrons were provided through the fifth ligand (histidine nitrogen) and the delocalized π-orbitals of the porphyrin. She then suggests (*215*) that since the electrons lost to the dioxygen are from the metal $d(\pi)$ orbitals, any factor which increases the π electron donation (from porphyrin to metal) will increase the stability of the metal-dioxygen system. *Halton's* theory (*215*) of two electron transfer contains two main features: first, the iron exists as a d^6 species in oxyhemoglobin (after the metal to O_2 electron transfer), and second, the dioxygen exists as an oxyanion. So although *Halton* does calculate a large negative charge on the dioxygen, in partial agreement with *Weiss*, she deduces that the electron transfer is compensated by a transfer from the protein to the iron, which therefore remains in the ferrous state. *Halton* has also carried out a similar study on oxycoboglobin (*216*) and it appears that cobalt is capable of forming stable monomeric adducts without the transfer of two electrons to the $Co \cdot O_2$ moitey, as deduced for oxyhemoglobin. This might explain the comparative scarcity of iron-dioxygen complexes. Oxycoboglobin has an unpaired electron which from ESR data (*56*) has been estimated to have 20% metal character; this in good agreement with the value of 25% predicted (*216*) by *Halton.* However, most authors have assumed that the remaining electron density is situated on the dioxygen, leading to a $Co(III) \cdot O_2^-$ configuration; inferred from ESR studies for various cobalt(II) dioxygen adducts (*68, 138, 139, 147, 217*), and predicted (*146*) for Co(acacen)L(O_2), where L = none, H_2O, imidazole, CN or CO, from an ab initio minimal basis set MO calculation. *Halton* deduces (*216*) that the cobalt is in a d^5 configuration, with the unpaired electron having 25% $3d(\pi)$, 20% O_2 and 55% porphyrin π character. The movement (*198*) of the metal ion from 0.08 nm below the porphyrin plane to a planar arrangement during oxygenation of hemoglobin, being sufficient to trigger the cooperative effects of haem-haem interaction, is consistent with *Halton's* proposals (*215, 216*) concerning electron donation from the porphyrin π orbitals to the metal orbitals in oxyhemoglobin. Further calculations (*218*) on the nitrosyl and carbon monoxide adducts of hemoglobin and coboglobin appear to vindicate the simplified model for the porphyrin ring which was employed in the previous calculations. Another calculation (*219*) also finds an energy minimum for the II(*P*) orientation for dioxygen in HbO_2. Although more recent results (*222*) suggest that it would be unwise to attach too much significance to this energy minimum. *Heitler-London* calculations (*223, 224*) indicate that a full explanation of the diamagnetism of HbO_2 and MbO_2 will require a greater emphasis on the role of paramagnetic electronic configurations, and this can be achieved by inclusion of configurational interaction in the calculations (*222*). The physics of hemoglobin, including an account of the applications of MO theory and wave-mechanical ESR and magnetic theory, has been reviewed in detail elsewhere (*225*). Assuming the II(*G*) orientation for the dioxygen ligand in HbO_2 and MbO_2, and using the MO results of *Zerner et al.* (*220*); *Weissbluth et al.* (*221*) calculated $\Delta E_Q = 2.1$ mm/sec and $V_{zz} < 0$, which are in agreement (*222*) with the experimental *Mössbauer* results (*227*), although a recent theoretical study (*234*), which assumes 'end-on' dioxygen bonding, finds that the ΔE_Q value is very sensitive to changes in the Fe—O—O angle and torsional angle (about the 1-Me-Imidazole-iron-dioxygen grouping present in the model system used in the calculation). An extensive review has been published recently (*228*) on the status of MO calculations on porphyrins and their complexes, however, this is mainly concerned with calculations designed to interpret the optical absorption spectra of these highly coloured complexes and relatively few calculations are performed on oxygen-carrying metal porphyrins, for the purpose of elucidating the nature of the dioxygen bonding, although some (*215, 216, 219, 220, 229*) have already been discussed. A recent X-ray analysis (*230*) of a manganese(II) porphyrin reports that the manganese atom lies in an out-of-plane position, and this in in accord with theoretical predictions (*229*). An IR difference spectral study (*231*) on packed human erythrocytes treated with $^{16}O/^{18}O$ or $^{16}O/CO$ assigns a unique band at 1107 cm^{-1} to the ^{16}O—^{16}O stretching mode and this is thought to imply a II(*P*) structure for dioxygen in oxyhemoglobin with an O—O bond order of 1.5. However, the

$^{16}O-^{16}O$ stretch has been assigned to a band at 1385 cm^{-1} in the spectrum of an oxygenated synthetic iron(II) porphyrin complex (*240*).

We shall return to the discussion of the orientation and electronic structure of the dioxygen ligand during the next section; which is primarily concerned with recent advances in the synthesis and physical characterization (particularly X-ray analysis) of oxygen-carrying metal porphyrin model compounds.

C. Synthetic Metal Porphyrins of Current Interest

It should be clear from Section IV. B that a major difficulty involved in preparing monomeric iron-dioxygen adducts is the prevention of bimolecular termination reactions, leading via autoxidation to the formation of a μ-oxo dimer, thus

$$Fe(II) + O_2 \rightleftharpoons Fe{-}O_2 \xrightarrow{Fe(II)} Fe(III){-}O{-}Fe(III)\,.$$

We saw previously that a major factor in inhibiting the bimolecular termination reaction was the presence of sufficiently bulky ligands so that a monomeric dioxygen adduct could be isolated (*135*). A number of synthetic metal porphyrins (*239*) have been prepared recently which satisfy the above requirement, and bind molecular oxygen; we shall now proceed to discuss these.

Collman et al. have reported the preparation (*235, 232*) and complete X-ray analysis (*237*) of the monomeric dioxygen adduct of ferrous "picket-fence" prophyrin; binding O_2 reversibly at room temperature in solution and in the solid state. Solid state thermodynamic measurements have been carried out (*282*) on the uptake of O_2 by Fe($\alpha,\alpha,\alpha,\alpha$-TpivPP) (1-Me Im), "picket-fence" porphyrin. Reversible oxygenation in both Mb and Hb probably (*235*) results from a five co-ordinate high spin iron(II) porphyrin fixed within a hydrophobic cavity. *Collman et al.* condensed *o*-nitrobenzaldehyde and pyrrole to give meso-tetra (*o*-nitrophenyl) porphyrin, which was then reduced to give mesotetra (*o*-aminophenyl) porphyrin, H_2TamPP. This product undergoes phenyl-type atropisomerism (*238*); however, the $\alpha,\alpha,\alpha,\alpha$-$H_2$TamPP isomer can be separated using thin layer silica gel chromatography, being the slowest moving isomer (*235*). Then this configuration was frozen by treating the amine with pivaloyl chloride to give the amide, $\alpha,\alpha,\alpha,\alpha$-$H_2$TpivPP, shown in Fig. 14, from which the iron(II) complex was isolated (*235*).

To prevent co-ordination of O_2 to the unhindered side of the iron, a benzene solution of Fe($\alpha,\alpha,\alpha,\alpha$-TpivPP) was prepared (*235*) under nitrogen in the presence of a small excess of l-methyl imidazole. This could then be reversibly oxygenated at 25 °C, but in the absence of excess ligand gradual oxidation takes place; presumably via dioxygen co-ordination at the open side of the porphyrin followed by bimolecular termination and autoxidation. Two other ligands obtained by treatment of $\alpha,\alpha,\alpha,\alpha$-$H_2$TamPP with p-toluenesulphonyl chloride and 1,3-phthaloylchloride, respectively, gave ferrous complexes that underwent irreversible oxidation in solutions containing an excess of axial ligand, on exposure to O_2, despite the bulky nature of the porphyrins (*232*). This was attributed (*232*) to protonation of the

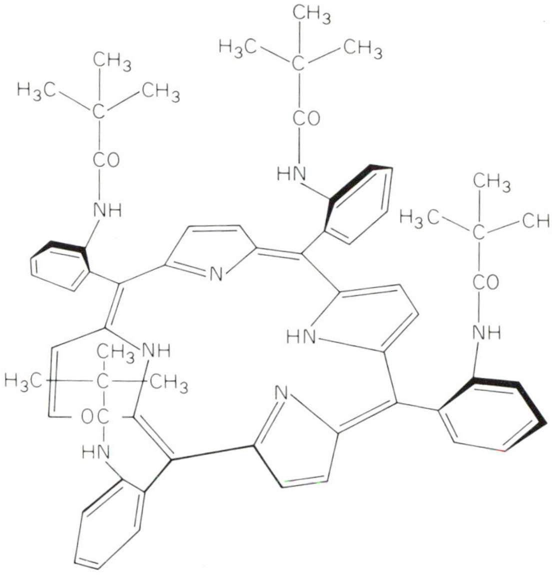

Fig. 14. "Picket fence" porphyrin; $\alpha,\alpha,\alpha,\alpha$-$H_2$TpivPP (Ref. *235*)

co-ordinated dioxygen by the acidic amide protons in these complexes, leading to haem oxidation. The four co-ordinate ferrous "picket-fence" porphyrin, Fe($\alpha,\alpha,\alpha,\alpha$-TpivPP), reacts with the strong field ligands (L = 1-MeIm, 1-*n*-BuIm, 1-trityl Im, 4-*t*-BuIm) in benzene solution to form diamagnetic six co-ordinate complexes Fe($\alpha,\alpha,\alpha,\alpha$-TpivPP)$L_2$; however, it is probable (*232*) that the binding constant of an imidazole on the "picket-fence" side of the porphyrin is lower than that on the open side, thus facilitating five co-ordination and therefore yielding an active "protected" co-ordination site for the dioxygen ligand; the equilibria involved being illustrated in Fig. 15.

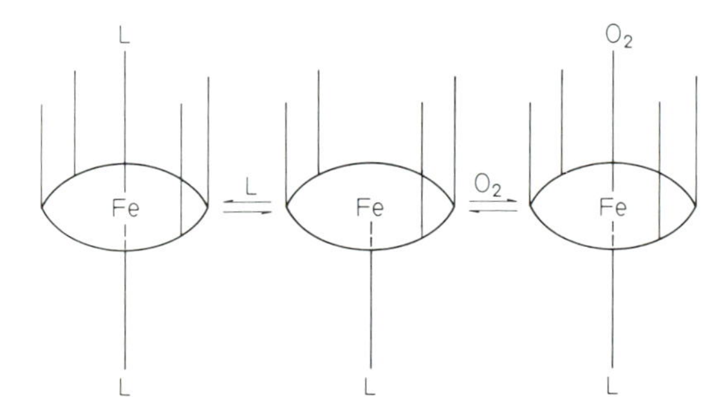

Fig. 15. Equilibria involved in oxygenation of Fe($\alpha,\alpha,\alpha,\alpha$-TpivPP) L_2 (Ref. *232*)

When L = 4-t-BuIm or 1-trityl Im, the dioxygen adduct could not be isolated, the less soluble six-co-ordinate species being preferentially precipitated (*232*). However, when L = 1-MeIm or 1-n-BuIm, the diamagnetic dioxygen adduct, Fe(α,α,α,α-TpivPP) (L) (O_2), could be isolated successfully by recrystallization under nitrogen, from a benzene solution containing excess ligand, L. An X-ray analysis of the dioxygen adduct Fe(α,α,α,α-TpivPP) (1-MeIm) (O_2) has been reported (*237*); the dioxygen is co-ordinated in the II(P) orientation with an Fe–O–O angle of 136°, and bond lengths Fe–O = 0.175 nm, Fe–N = 0.207 nm and O–O = 0.125 nm. The iron lies in the plane of the porphyrin (*237*), as previously reported for HbO_2 (*198*). The 1-MeIm is two-fold disordered with respect to the orientation of the N-methyl group, and the Fe–O–O plane is four-way statistically disordered bisecting the N–Fe–N right angles formed between the iron and porphyrin nitrogen atoms (*232*). Hence the II(P) dioxygen ligand is co-ordinated with the Fe–O–O plane either parallel or perpendicular to the plane of the trans axial imidazole ligand. The O–O bond length of 0.125 nm is suggestive of a superoxide species, however, a low temperature infra-red spectral study (*240*) finds a band at 1385 cm^{-1}, assigned to the O_2 stretch, which indicates substantial double bond character in the dioxygen but sufficient lowering of the O_2 stretching frequency to suggest moderate $d(\pi)$ to $p(\pi)$ back bonding. The iron is thought (*240*) to be in a low-spin Fe(II) state, with the dioxygen present as co-ordinated singlet O_2, which is largely consistent with Halton's proposals (*215, 216*). The dioxygen complex can be deoxygenated in the solid phase by placing a powdered sample under vacuum, which can be oxygenated again on exposure to O_2. Also, the dioxygen adduct (where one THF is a crystal solvate) Fe(α,α,α,α-TpivPP) $(THF)_2(O_2)$ has been isolated (*241*) by oxygenation of crystals of the complex Fe(α,α,α,α-TpivPP) $(THF)_2$; the dioxygen adduct could not be isolated from oxygenated solutions of the THF complex, this yielded the oxidized μ-oxo product (Fe(α,α,α,α-TpivPP))$_2$O, mostly. The novel feature of the THF complex is its paramagnetism, which could not be explained by any artifact of the experiment (*241*). The work on "picket-fence" porphyrin represents a substantial step forward in our understanding of HbO_2 and MbO_2, within the context of this review, being especially important since it involves a model porphyrin that can be oxygenated at 25 °C, not requiring low temperatures (*135, 242*) to retard irreversible oxidation, and yields a dioxygen adduct which has been isolated and structurally characterised by X-ray analysis. Many reports of ferrous complexes which reversibly bind O_2 rest on the observation of a single physical parameter, often a change in the visible spectrum, as a means of following the reversible uptake of O_2. But in view of reported (*243, 244*) facile ligand redox reactions of iron complexes, claims of reversible oxygenation which rely purely on spectral evidence must be treated with a good deal of caution (*235*). The four-way statistical disordering of the terminal oxygen atom in Fe(α,α,α,α-TpivPP) (1-MeIm) (O_2), elucidated (*237*) by X-ray analysis, may (*232*) account for the temperature dependence of ΔE_Q, the quadrupole splitting, reported (*232*) from a Mössbauer analysis of this complex, and which is also observed (*227, 245*) in the Mössbauer spectrum of HbO_2. This is consistent with a theoretical study (*234*) on a model oxy Hb compound which shows that ΔE_Q is very sensitive to changes in orientation; the dioxygen was assumed (*234*) to be bonded 'end-on', but both the Fe–O–O angle, and the direction of the projection of the O–O bond on the porphyrin plane, were varied over wide ranges. The ferrous complex formed with the tetraphenylporphyrin ligand (in the absence of "picket-fence" or other bulky substituents) does react reversibly with O_2 at low temperatures (–78 °C) in the presence of excess axial base (*246*),

$$Fe(B)_2TPP \overset{O_2}{\rightleftharpoons} Fe(B)(O_2)TPP + B .$$

Where B = pyridine, piperidine or 1-methylimidazole, in methylene chloride solution, but under normal conditions rapid irreversible autoxidation takes place (*232*) leading to the formation of the well characterised (*247, 248*) μ-oxo product, (TPP)Fe(III)–O–Fe(III)(TPP); and since the rate of oxidation decreases (*249, 250*) with increasing excess of axial base, B, it follows (*232, 251*) that a five co-ordinate species, Fe(II) (Base)TPP, is probably involved as an intermediate; which can then undergo a bimolecular termination reaction with Fe(II) (Base)O_2TPP, followed by autoxidation. Firstly (*251*),

$$\left.\begin{array}{l} Fe(II)(B)_2TPP \rightleftharpoons Fe(II)(B)TPP + B \\ Fe(II)(B)TPP \xrightarrow{O_2} Fe(II)(B)(O_2)TPP \end{array}\right\} \text{Reaction Scheme I}$$

which will be followed (*232*) under normal conditions by

$$[TPP(B)Fe(II)] + (O_2)Fe(II)(B)TPP \xrightarrow{\text{bimolecular termination}}$$
$$TPP(B)Fe(II)\text{–}O\text{–}O\text{–}Fe(II)(B)TPP \xrightarrow{\text{irreversible autoxidation}}$$
$$TPPFe(III)\text{–}O\text{–}Fe(III)TPP .$$

When B = imidazole, irreversible oxidation occurs (*261*) at –78 °C in a toluene solution of Fe(II)TPP with B. The kinetics and thermodynamics of Reaction Scheme I have been studied recently (*251*), and when there is a large excess of O_2 (no added base) the forward rate constant is independent of the O_2 concentration, whereas when there is an excess of base the backward (deoxygenation) rate constant is independent of the base concentration; both these observations support the earlier contention (*232*) that the five co-ordinate species Fe(II) (B)TPP is involved as an intermediate (*251*). In the case of the intermediate Fe(TPP) (1-MeIm), the rate of reaction with O_2 and 1-MeIm is equal (*251*), so that the 1-MeIm co-ordinates to the iron in competition with the dioxygen, whereas in naturally occurring hemoproteins the distal-imidazole grouping is structurally positioned (*197, 255*) in such a way that it cannot co-ordinate to the iron in competition with the dioxygen (*251*). The extent of dioxygen adduct formation at – 78° by Fe(py)$_2$TPP in various solvents, under 1 atm of O_2, varies (*251*) according to the solvent as follows (*246*)

methylene chloride	>	ethyl ether	>	toluene
9.08		4.34		2.39

the values shown are for the dielectric constants for these solvents at 20 °C; so that the equilibrium constant for the formation of Fe(py) (O_2)TPP will increase as the dielectric strength of the solvent increases (*251*), as found (*252*) for cobalt porphyrin dioxygen adducts.

A silica gel containing 3-imidazole propyl groups bonded to surface atoms of silicon has been prepared (*253*), and was treated (*254*) with a solution of $Fe(TPP)L_2$, where L = py or pip, where upon the iron(II) porphyrin became attached to the imidazole groups on the surface of the gel,

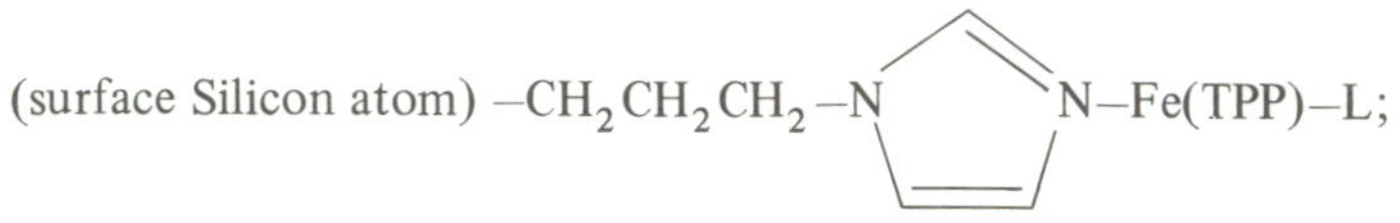

(surface Silicon atom) $-CH_2CH_2CH_2-$N [imidazole ring] N–Fe(TPP)–L;

on heating at 250 °C in a flow of helium, the remaining ligand, L, was removed giving co-ordinatively unsaturated (five-co-ordinate) surface iron atoms, which were able (*254*) to reversibly bind O_2; the dioxygen is chemisorbed weakly at 0 °C, strongly at –78 °C and irreversibly at –127 °C. In the above study (*254*) dimerization to form the μ-oxo product, did not occur since the support was rigid enough to prevent this; in contrast to a previously investigated (*233*) modified polystyrene surface containing imidazole ligands. It is interesting that a toluene solution of manganese(II) (TPP), when treated (*230*) with ligand L(1-MeIm, 2-MeIm, or py), in large excess, does not form six-co-ordinate species as did the analogous iron porphyrin; only the five-co-ordinate species Mn(TPP) (L) were formed; the manganese being in a high-spin d^5 configuration. An X-ray analysis (*230*) of Mn(TPP) (1-MeIm) confirms the existence of five-co-ordinate manganese in this complex, and that the manganese atom lies out of the porphyrin plane, as predicted theoretically (*229*). Solutions of the four-co-ordinate Mn(TPP) complex are oxidised on exposure to O_2, whereas Mn(TPP)L or Mn($\alpha,\alpha,\alpha,\alpha$-TpivPP)L are inactive to O_2 although irreversible oxidation to Mn(III) products does occur ultimately (*230*); Mn(II)Hb is oxidized to Mn(III)Hb on exposure to O_2 (*256*). The inability of dioxygen to produce a change from high to low spin in the hypothetical complex $Mn(O_2)$ (TPP)L, precludes its existence (*230*). However, largely reversible oxygenation of a toluene-THF solution of Mn(TPP) occurs at –90 °C to produce (*230*) the five-co-ordinate species $Mn(O_2)$ (TPP), and the oxygenation of a toluene solution of Mn(TPP) (py) at –79 °C has been shown (*257*), by a temperature-dependent ESR analysis of the reaction, to proceed as follows

$$(TPP)MnPy + O_2 \rightleftharpoons (TPP)MnO_2 + Py$$

and not via the six-co-ordinate species Mn(Py) (O_2) (TPP). The dioxygen adduct has been formulated (*257*) as the peroxo complex Mn(IV) (TPP) (O_2^{2-}), from the ESR spectrum, and a II(*G*) orientation for the dioxygen ligand (rather than the II(*P*) orientation found for iron and cobalt porphyrins) is considered (*257*) most favourable in this case. The ESR and electronic spectra for Co(II) (O_2)TPP have been obtained (*258*) and a molecular orbital model has been proposed to explain the structure and bonding of low spin cobalt(II) and iron(II) adducts formed with diatomic molecules such as dioxygen. The metal d_{z^2} orbital is singly occupied in cobalt(II)- and empty in iron(II)-porphyrin dioxygen complexes – on the basis of the above mentioned (*258*) MO scheme – and this ultimately leads to greater electron density localisation on the dioxygen in the case of the $Co-O_2$ moitey, as compared to the $Fe-O_2$ moitey; the iron-porphyrin dioxygen adducts are then formulated as singlet O_2 binding Fe(II), whereas the cobalt-porphyrin dioxygen adducts are assigned the familiar electronic structure Co(III)$-O_2^-$ for the cobalt-dioxygen moitey (*258*).

A study (*259*) of the reversibly formed dioxygen adducts formed with natural protoheme IX in solution, with either imidazole or a primary amine ligand, concludes that the π-bonding ability of the axial ligand, such as imidazole, is not an essential requirement for reversible O_2 uptake; the ligand t-butylamine is unable to π-bond with the heme, and yet this complex forms a reversible dioxygen adduct; the chemical and physical conditions (to which we have already eluded) are thought (*259*) to be more important in determining dioxygen uptake properties.

By direct condensation of suitable tetraaldehydes with pyrolle, workers (*236*) have succeeded in synthesizing "capped" porphyrins; a particular one is shown in Fig. 16.

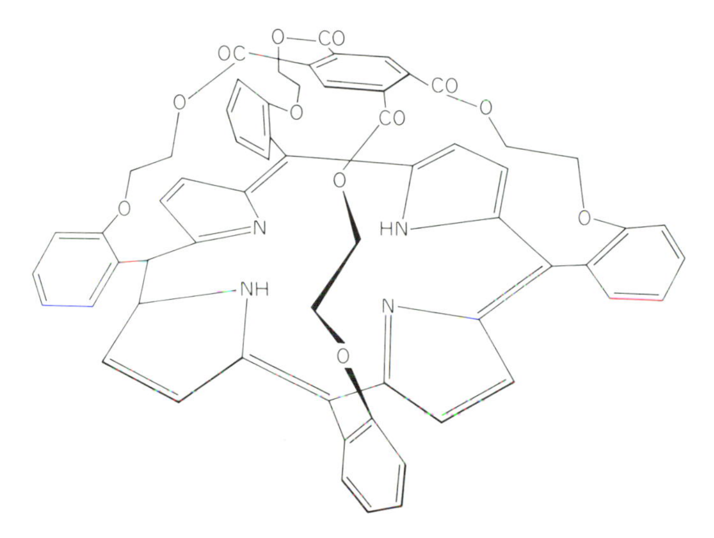

Fig. 16. "Capped" porphyrin ligand

Ferrous "capped" porphyrin has been prepared (*260*), and in solutions containing a high concentration of co-ordinating base, five-co-ordinate iron will result (thus inhibiting oxygenation of the "open" side of the prophyrin), and so reversible oxygenation can occur at the "protected" co-ordination site. In the absence of co-ordinating ligand the ferrous "capped" porphyrin is instantly autoxidized (*260*) on exposure to O_2, via oxygenation of the "open" unprotected face followed by bimolecular termination; as discussed previously for other samples. The rate of autoxidation is therefore lowered (*260*) by an increase in either the concentration of the ligand, or by using a better co-ordinating ligand; a 5% concentration of 1-MeIm in benzene gives rise to a 5 hr lifetime (25°), whereas a 5% pyridine, in benzene, solution leads to rapid autoxidation.

A series of synthetic heme-compounds have been prepared (*242, 262, 263*) which have the promimal nitrogen base covalently attached to the side-chain on the porphyrin, in a similar manner to that found in myoglobin. Reversible oxygenation was found (*242*) to take place in solution (low temperatures being necessary to retard irreversible oxidation), as well as in the solid state (*262*). Increased solvent polarity was found (*264*) to favour oxygenation for these heme-base complexes, implying a highly dipolar Fe–OO bond; as was found (*252*) for the dioxygen complexes formed by cobalt porphyrins, and which led to a $Co(III) \cdot O_2^-$ formulation for the cobalt-dioxygen moitey. It was concluded (*261*) contrary to earlier reports (*242, 263*), that there is no requirement for a covalently attached neighbouring group effect for reversible oxygenation to occur, since iron(II) complexes of mesoporphyrin IX dimethyl ester reversibly bind O_2 at –50 °C in the presence of 1-methylimidazole (*261*), which is clearly not covalently bonded to the porphyrin through a side-chain, as in the myoglobin model compounds (*242*). However, two myoglobin model compounds – monopyridine propanol meso heme iron(II) and pyrroheme-*N*-(3-(1-imidazolyl)-propyl)-amide iron(II) – have been investigated (*265*), and only the latter complex binds dioxygen; this is attributed (*265*) to the effect of the proximal imidazole which has a larger π-basicity than pyridine (the proximal base in the first complex); however, both complexes bind CO strongly, since CO-binding will tend to be affected mainly by changes in sigma-type basicity, and therefore the change from pyridine to imidazole as the proximal base in the above complexes does not radically alter their CO binding ability.

The complex Bis(L) ferroprotoporphyrin dimethyl ester, where L = imidazole or 1-butylimidazole, combines with O_2 reversibly at – 45 °C in DMF, but the presence of the butyl group when L = 1-butylimidazole results in greater stability toward oxidation (*266*); however, in the absence of stronger ligands, L, a five co-ordinate complex is formed (either with DMF itself or dimethylamine, a possible impurity) which also combines reversibly with O_2 at – 45 °C (*266*); in all cases oxidation occurs at 20 °C. A recently prepared (*281*) myoglobin model compound (containing a covalently attached proximal base) has been shown to exist almost entirely as a five co-ordinate species in aqueous solution, even though there is no hydrophobic pocket surrounding the sixth site; this is attributed (*281*) to an inherent property of the heme molecule.

Cobaltohemoglobin has been prepared (*271*), and it has been shown (*267*) that it undergoes a quaternary structure transition upon oxygenation, as does FeHb, and consequently exhibits Bohr and phosphate effects. The binding constant of dioxygen to both "whale" and "horse skeletal muscle" cobaltomyoglobin is 300X greater than for the N-methylimidazole complex of cobalt protoporphyrin IX; for the horse-CoMb the binding enhancement is entirely entropic (300-fold), whereas for whale-CoMb the binding enhancement is 20-fold enthalpic and 15-fold entropic (*268*). *Walker* (*269*) studied the equilibrium

$$Co \cdot P(\mathrm{B}) + O_2 \rightleftharpoons Co \cdot P(\mathrm{B})(O_2)$$

where *CoP* = Co(II) Porphyrin and (B) = amine, and no correlation could be found between the sigma-donor strength of (B) and the stability of the $Co \cdot O_2$ moitey, although back donation from cobalt to (B) seems to be significant. It was also found that the $Co \cdot P\,(O_2)$ complex was oxidatively unstable, which indicates that the amine (B) plays an important role in the formation of a stable dioxygen adduct (*269*); though it seems that the activation of Co(II) porphyrins toward O_2, and the ESR parameters of the dioxygen adducts (*143*), are much less sensitive to changes in the axial bases than are Schiff's bases and DMG with alkyl groups (*269, 270*). There has been some controversy regarding the measurement and analysis of thermodynamic data for the binding of dioxygen to cobalt(II) protoporphyrin IX dimethyl ester (*272, 273, 274, 275, 276*). For instance, the enthalpy of dioxygen interaction was reported (*273*) to be −9.2 ± 0.6 kcal/mole, and yet more recent measurements (*277*) yield a value of − 7.80 kcal/mole; for the pyridine complex of cobalt(II) protoporphyrin IX dimethyl ester. However, it has been concluded (*274*) that the π-donor properties of the axial ligand play a major role in determining O_2 binding to the cobalt in the corresponding protoporphyrin IX dimethyl ester complex. Recently, models for the naturally occurring O_2 binding molecule cytochrome P-450 have been synthesized (*278–280*); the dioxygen, co-ordinated to the naturally occurring cytochrome oxidase, is reduced in the terminal step of the electron transport chain (*232*), and further investigation of these model compounds (*278–280*) should lead to a greater understanding of this process which may also involve reaction at the heme group (*232*). In the oxy-P-450 model species there is significant π character in both the iron-O_2 and iron-sulphur bonds, therefore changes in the geometry of either the dioxygen or mercaptide axial ligands (in a model theoretical complex studied), will have an affect on the EFG (*234*). However, the work on cytochrome-P-450 (and the synthetic model compounds) is still at an early stage of development; and in any case, this system involves reduction of the co-ordinated dioxygen in the oxy compound, and is therefore somewhat outside the scope of this review, which is primarily concerned with truly reversible "non-reactive" dioxygen uptake.

VI. Concluding Remarks

We have attempted to discuss dioxygen complexes in terms of the orientation of the dioxygen ligand (I, II(*P*) or II(*G*) adducts), and the factors affecting the reversibility and stability of these complexes. Firstly, a consideration of the relative energies of the metal and dioxygen valence orbitals and a means of estimating the metal-dioxygen orbital-overlap, is required in order to provide an "orientation criterion" capable of predicting whether a II(*P*) or II(*G*) orientation is preferred; see Section V. B. If a II(*G*) orientation is favoured, we can obtain a 'reversible' dioxygen adduct, in principle, by an appropriate choice of ligand(s); see Section IV. C. Whereas reversible II(*P*) adducts can be isolated so long as bimolecular termination, which leads to autoxidation for iron(II) complexes, does not take place; see Sections IV. B and V. C. Reversible type I dioxygen adducts, stable to irreversible oxidation, of cobalt(II) complexes are commonly formed in solution via bimolecular termination involving monomeric II(*P*) type radicals (see Section IV. A), and in aqueous media will invariably contain μ-hydroxo bridging in addition to the dioxygen bridging, providing appropriate vacant sites are available on the cobalt atoms; if (O_2, OH) dibridging is present, addition of acid is generally required to destroy the hydroxo bridge before O_2 can be recovered from the adduct. Finally, recent model synthetic iron(II) porphyrins have been synthesized (see Section V. C) and, in some cases, are well characterized. For reversible oxygenation to be feasible in solution *at room temperature*, it has been found necessary to obtain five-co-ordinate metal species with the vacant co-ordination site protected by bulky substituents on the porphyrin ligand and the axial ligand being present in solution in excess in order that bimolecular termination (followed by irreversible autoxidation, leading to the formation of an iron(III) product, probably μ-oxo) could be prevented. We have discussed the electronic structure of the dioxygen-metal grouping for type I, II(*P*) and II(*G*) adducts (see Sections IV and V. B) and have also considered correlations between changes in ligand with its effect on the stability and reversibility of these dioxygen adducts. Clearly, there is a great deal more experimental work that must be carried out in this field; on synthetic model porphyrin complexes in particular. Although both the theoretical and experimental areas are anticipated to make significant progress in the near future, a definitive theoretical treatment of the subject, capable of making quantitative testable predictions, has not yet been developed.

Acknowledgment: We wish to thank the Science Research Council for the Award of a Research Studentship (to RWE).

References

1. *Valentine, J.S.:* Chem. Rev. *73*, 235 (1973).
2. *Hughes, M.N.:* The Inorganic Chemistry of Biological Processes. London: Wiley 1972.
3. *Fallab, S.:* Angew. Chem. Int. Ed. Engl. *6*, 496 (1967).
4. *Wang, J.H., Brinigar, W.S.:* Proc. Nat. Acad. Sci. *46*, 958 (1960).
5. *Nakamura, T.:* Biochim. Biophysica Acta *42*, 499 (1960).
6. *Latimer, W.M.:* The Oxidation States of the Elements in Aqueous Solution. New York: Prentice-Hall 1952.
7. *Ochiai, E.I.:* J. Inorg. Nucl. Chem. *37*, 1503 (1975).
8. *George, P.:* Oxidases and Related Redox Systems. New York: Wiley 1965.
9. *Pfeiffer, P., Tsumaki, T., Breith, E., Lubbe, E.:* Ann. Chem. Liebigs *503*, 84 (1933).
10. *Tsumaki, T.:* Bull. Chem. Soc. Japan *13*, 252 (1938).
11. *Bailes, R.H., Calvin, M.:* J. Amer. Chem. Soc. *69*, 1886 (1947).
12. *Calvin, M., Bailes, R.H., Wilmarth, W.K.:* J. Amer. Chem. Soc. *68*, 2254 (1946).
13. *Barkelew, C.H., Calvin, M.:* J. Amer. Chem. Soc. *68*, 2257 (1946).
14. *Wilmarth, W.K., Aranoff, S., Calvin, M.:* J. Amer. Chem. Soc. *68*, 2263 (1946).
15. *Calvin, M., Barkelew, C.H.:* J. Amer. Chem. Soc. *68*, 2267 (1946).
16. *Hughes, E.W., Wilmarth, E.K., Calvin, M.:* J. Amer. Chem. Soc. *68*, 2273 (1946).
17. *Martell, A.E., Calvin, M.:* Chemistry of the Metal Chelate Compounds. New York: Prentice-Hall 1953.
18. *Ueno, K., Martell, A.E.:* J. Chem. Phys. *60*, 1270 (1956).
19. *Vogt, L.H., Faigenbaum, H.M., Wiberley, S.E.:* Chem. Rev. *63*, 269 (1963).
20. *Ferroni, E., Ficalbi, A.:* Gazzeta *89*, 750 (1959).
21. *Yamada, S., Nishikawa, H.:* Bull. Chem. Soc. Japan *37*, 8 (1964).
22. *Yamada, S., Nishikawa, H., Yoshida, E.:* Proc. Japan Academy *40*, 211 (1964).
23. *Figgis, B.M., Nyholm, R.S.:* J. Chem. Soc. *1954*, 12.
24. *West, B.O.:* J. Chem. Soc. *1954*, 395.
25. *Bayer, E., Schretzmann, P.:* Struct. Bonding *2*, 181 (1967).
26. *Diehl, H., Henn, J.:* Iowa State Coll. J. Sci. *23*, 273 (1949).
27. *Diehl, H., Hach, C.C., Harrison, G.C., Liggett, L.M., Chao, T.S.:* Iowa State Coll. J. Sci. *21*, 326, (1947).
28. *Stewart, R.F., Estep, P.A., Sebastian, J.J.S.:* U.S. Bur. Mines Inform. Circ. No. 7906 (1959).
29. *Hewlett, P.C., Larkworthy, L.F.:* J. Chem. Soc. *1965*, 882.
30. *Pauling, L.:* Hemoglobin, Sir Joseph Barcroft Memorial Symposium. London: Butterworths 1949.
31. *Pauling, L.:* Nature *203*, 182 (1964).
32. *Griffith, J.S.:* Proc. Roy. Soc. *A 235*, 23 (1956).
33. *Hearon, J.Z., Burk, D., Schade, A.L.:* J. Natl. Cancer Inst. *9*, 337 (1949).
34. *Hearon, J.Z.:* J. Natl. Cancer Inst. *9*, 1 (1948).
35. *Caglioti, V., Silvestroni, P., Furlani, C.:* J. Inorg. Nucl. Chem. *13*, 95 (1960).
36. *Sano, Y., Tanabe, H.:* J. Inorg. Nucl. Chem. *25*, 11 (1963).
37. *Gilbert, J.B., Otey, M.C., Price, V.E.:* J. Biol. Chem. *190*, 377 (1951).
38. *Tanford, C., Kirk, jun., D.C., Chantooni, jun., M.K.:* J. Amer. Chem. Soc. *76*, 5325 (1954).
39. *Miller, G.W., Li, N.C.:* Trans. Faraday Soc. *57*, 2041 (1961).
40. *Smith, E.L.:* J. Biol. Chem. *173*, 571 (1948).
41. *Simplicico, J., Wilkins, R.G.:* J. Amer. Chem. Soc. *89*, 6092 (1967).
42. *Earnshaw, A., Larkworthy, L.F.:* Nature *192*, 1068 (1961).
43. *Watters, K.L., Wilkins, R.G.:* Inorg, Chem. *13*, 752 (1974).
44. *Paniago, E.B., Weatherburn, D.C., Margerum, D.W.:* J.C.S. Chem. Commun. *1971*, 1427.
45. *McDonald, C.C., Phillips, W.D.:* J. Amer. Chem. Soc. *85*, 3736 (1963).
46. *Foong, S.W., Miller, J.D., Oliver, F.D.:* J. Chem. Soc. (A) *1969*, 2847.
47. *Gillard, R.D., Spencer, A.:* J. Chem. Soc. (A) *1969*, 2718.

48. *Michailidis, M. S., Martin, R. B.:* J. Amer. Chem. Soc. *91*, 4683 (1969).
49. *Gillard, R. D., Spencer, A.:* Discuss. Faraday Soc. *46*, 213 (1968).
50. *Gillard, R. D., McKenzie, E. D., Mason, R., Robertson, G. B.:* Co-ordination Chem. Rev. *1*, 263 (1966).
51. *Jonassen, H. B., Schaafsma, A., Westerman, L.:* J. Phys. Chem. *62*, 1022 (1958).
52. *Drake, J. F., Williams, R. J. P.:* Nature *182*, 1084 (1958).
53. *Cowan, M. J., Drake, J. M. F., Williams, R. J. P.:* Discuss. Faraday Soc. *27*, 217 (1959).
54. *Davies, R. C.:* Ph. D. Thesis, Wadham College, Oxford, 1963.
55. *Selbin, J., Junkin, J. H.:* J. Amer. Chem. Soc. *82*, 1057 (1960).
56. *Schrauzer, G. N., Lee, L. P.:* J. Amer. Chem. Soc. *92*, 1551 (1970).
57. *Hayward, G. C., Hill, H. A. O., Pratt, J. M., Vanston, N. J., Williams, R. J. P.:* J. Chem. Soc. *1965*, 6485.
58. *Bayston, J. H., King, N. K., Looney, F. D., Winfield, M. E.:* J. Amer. Chem. Soc. *91*, 2775 (1969).
59. *Crumblies, A. L., Basolo, F.:* Science *164*, 1168 (1969).
60. *Werner, A., Mylius, A.:* Z. Anorg. Chem. *16*, 245 (1898).
61. *Fallab, S.:* Z. nat-med. Grundlagenforschung *2*, 2201 (1965).
62. *Fremy, E.:* Ann. Chim. Phys. *(3) 25*, 257 (1852).
63. *Vortmann, G.:* Monatsch. *6*, 404 (1885).
64. *Bjerrum, J.:* Metal Ammine Formation in Aqueous Solution. Copenhagen: P. Haase and Son 1941.
65. *Lewis, W. B., Corgell, C. D., Irvine, jun., J. W.:* J. Chem. Soc. *1949*, S 386.
66. *Biradar, N. S., Stranks, D. R., Vaidya, M. S.:* Trans. Faraday Soc. *58*, 2421 (1962).
67. *Mori, M., Weil, J. A., Ishiguro, M.:* J. Amer. Chem. Soc. *90*, 615 (1968).
68. *Yang, N. L., Oster, G.:* J. Amer. Chem. Soc. *92*, 5265 (1970).
69. *Marsh, R. E., Schaefer, W. P.:* Acta Cryst. *B 24*, 246 (1968).
70. *Schaefer, W. P., Marsh, R. E.:* J. Amer. Chem. Soc. *88*, 178 (1966).
71. *Schaefer, W. P.:* Inorg. Chem. *7*, 725 (1968).
72. *Marsh, R. E., Schaefer, W. P.:* Acta Cryst. *21*, 735 (1966).
73. *Simplicio, J., Wilkins, R. G.:* J. Amer. Chem. Soc. *91*, 1325 (1969).
74. *Sykes, A. G.:* Trans. Faraday Soc. *59*, 1325 (1963).
75. *Hoffman, A. B., Taube, H.:* Inorg. Chem. *7*, 1971 (1968).
76. *Wilkins, R. G.:* Adv. Chem. Ser. *No. 100*, 111 (1971).
77. *Choy, V. J., O'Conner, C. J.:* Co-ordination Chem. Rev. *9*, 145 (1972–73).
78. *Henrici-Olivé, G., Olivé, S.:* Angew. Chem. Int. Ed. Engl. *13*, 29 (1974).
79. *Sykes, A. G., Weil, J. A.:* Progr. Inorg. Chem. *13*, 1 (1970).
80. *Taube, H.:* J. Gen. Physiol *49 (No. 1, Pt. 2)*, 29 (1965/1966).
81. *Ochiai, E. I.:* J. Inorg. Nuclear Chem. *35*, 3375 (1973).
82. *McGarvey, B. R., Tepper, E. L.:* Inorg. Chem. *8*, 498 (1969).
83. *Yamada, S., Shimura, Y., Tsuchida, R.:* Bull. Chem. Soc. Japan *26*, 72 (1953).
84. *McLendon, G., MacMillan, D. T., Hari Haran, M., Martell, A. E.:* Inorg. Chem. *14*, 2322 (1975).
85. *Stadtherr, L. G., Prados, R., Martin, R. B.:* Inorg. Chem. *12*, 1814 (1973).
86. *Miller, F., Simplicio, J., Wilkins, R. G.:* J. Amer. Chem. Soc. *91*, 1962 (1969).
87 *Sillen, L. G., Martell, A. E., Ed.:* Stability Constants of Metal-Ion Complexes, Special Publication No. 17. London: The Chem. Soc. 1964.
88. *Fallab, S.:* Chimia *21*, 538 (1967).
89. Ibid. *23*, 177 (1969).
90. *Miller, F., Wilkins, R. G.:* J. Amer. Chem. Soc. *92*, 2687 (1970).
91. *Nakon, R., Martell, A. E.:* J. Amer. Chem. Soc. *94*, 3026 (1972).
92. *Nakon, R., Martell, A. E.:* J. Inorg. Nucl. Chem. *34*, 1365 (1972).
93. *McLendon, G., Martell, A. E.:* J. C. S. Chem. Commun. *1975*, 223.
94. *Nakon, R., Martell, A. E.:* Inorg. Chem. *11*, 1002 (1972).
95. *Crook, E. M., Rabin, B. R.:* Biochem. J. *68*, 117 (1958).
96. *Huchital, D. H., Martell, A. E.:* Inorg, Chem. *13*, 2966 (1974).
97. *Powell, H. K. J., Nancollas, G. H.:* J. Amer. Chem. Soc. *94*, 2664 (1972).

98. *Kauffman, B.:* Co-ordination Chem. Rev. *9*, 339 (1972–73).
99. *Mori, M., Weil, J.A., Kinnaird, J.K.:* J. Chem. Phys. *71*, 103 (1967).
100. *Bayston, J.H., Looney, F.D., Winfield, M.E.:* Australian J. Chem. *16*. 557 (1963). – ibid., p. 954.
101. *Mori, M., Weil, J.A.:* J. Amer. Chem. Soc. *89*, 3732 (1967).
102. *Ebsworth, E.A.V., Weil, J.A.:* J. Phys. Chem. *63*, 1890 (1959).
103. *Weil, J.A., Kinnaird, J.K.:* J. Phys. Chem. *71*, 3341 (1967).
104. *Thompson. L.R., Wilmarth, W.K.:* J. Phys. Chem. *56*, 5 (1952).
105. *Garbett, K., Gillard, R.D.:* J. Chem. Soc. (A) *1968*, 1725.
106. *Fronczek, F.R., Schaefer, W.P.:* Inorg. Chim. Acta *9*, 143 (1974).
107. *Fronczek, F.R., Schaefer, W.P., Marsh, R.E.:* Inorg. Chem. *14*, 611 (1975).
108. *Miskowski, V.M., Robbins, J.L., Treitel, I.M., Gray, H.B.:* Inorg. Chem. *14*, 2318 (1975).
109. *Charles, R.G., Barnatt, S.:* J. Inorg. Nuclear Chem. *22*, 69 (1961).
110. *Sykes, A.G.:* Trans. Faraday Soc. *59*, 1325 (1963).
111. *Barrett, J.:* J. C. S. Chem. Commun. *1968*, 874.
112. *McLendon, G., Martell, A.E.:* Inorg. Chem, *14*, 1423 (1975).
113. *McKenzie, E.D.:* J. Chem. Soc. (A) *1969*, 1655.
114. *Hobday, M.D., Smith, T.D.:* Co-ordination Chem. Rev. *9*, 311 (1972–73).
115. *Calligaris, M., Nardin, G., Randaccio, L.:* J. C. S. Chem. Commun. *1969*, 763.
116. *Calligaris, M., Nardin, G., Randaccio, L., Ripamonti, A.:* J. Chem. Soc. (A) *1970*, 1069.
117. *Vannerberg, N.G.:* Acta Cryst. *18*, 449 (1965).
118. *Vannerberg, N.G., Brosset, C.:* Ibid. *16*, 247 (1963).
119. *Bruckner, S., Calligaris, M., Nardin, G., Randaccio, L.:* Acta Cryst. *B 25*, 1671 (1969).
120. *Delasi, R., Holt, S.L., Post, B.:* Inorg. Chem. *10*, 1498 (1971).
121. *Calligaris, M., Minichelli, D., Nardin, G., Randaccio, L.:* J. Chem. Soc. (A) *1970*, 2411.
122. *Calderazzo, F., Floriani, C., Salzmann, J.J.:* Inorg. Nucl. Chem. Letts. *2*, 379 (1966).
123. *Lewis, J., Mabbs, F.E., Richards, A.:* J. Chem. Soc. (A) *1967*, 1014.
124. *Gerloch, M., McKenzie, E.D., Towl, A.D.C.:* J. Chem. Soc. (A) *1969*, 2850.
125. *Earnshaw, A., King, E.A., Larkworthy, L.F.:* J. Chem. Soc. (A) *1968*, 1048.
126. *Lewis, J., Mabbs, F.E., Weigold, H.:* J. Chem. Soc. (A) *1968*, 1699.
127. *Yarino, T., Matsushita, Y., Masuda, I., Shinra, K.:* J. C. S. Chem. Commun. *1970*, 1317.
128. *Matsushita, T., Yarino, T.:* Bull. Chem. Soc. Japan *46*, 1712 (1973).
129. *Maslen, H.S., Waters, T.N.:* J. C. S. Chem. Commun. *1973*, 760.
130. *Hoffman, B.M., Diemente, D., Basolo, F.:* J. Amer. Chem. Soc. *92*, 61 (1970).
131. *Earnshaw, A., Hewlett, P.C., King, A., Larkworthy, L.F.:* J. Chem. Soc. (A) *1968*, 241.
132. *Burness, J.H., Dillard, J.G., Taylor, L.T.:* J. *Amer. Chem. Soc. 97*, 6080 (1975).
133. *Boucher, L.J., Coe, C.G.:* Inorg. Chem. *14*, 1289 (1975).
134. *Pauling, L., Coryell, C.D.:* Proc. Nat. Acad. Sci., Wash. *22*, 210 (1936).
135. *Baldwin, J.E., Huff, J.:* J. Amer. Chem. Soc. *95*, 5757 (1973).
136. *Baldwin, J.E., Holm, R.H., Harper, R.W., Huff, J., Koch, S., Truex, T.J.:* Inorg. Nucl. Chem. Letts. *8*, 393 (1972).
137. *Floriani, C., Calderazzo, F.:* J. Chem. Soc. (A) *1969*, 946.
138. *Diemente, D., Hoffman, B.M., Basolo, F.:* J. C. S. Chem. Commun. *1970*, 467.
139. *Crumbliss, A.L., Basolo, F.:* J. Amer. Chem. Soc. *92*, 55 (1970).
140. *Misono, A., Koda, S.:* Bull. Chem. Soc. Japan *42*, 3048 (1969).
141. *Koda, S., Misono, A., Uchida, Y.:* Ibid. *43*, 3143 (1970).
142. *Bussetto, C., Neri, C., Palladino, N., Perrotti, E.:* Inorg. Chim. Acta *5*, 129, (1971).
143. *Walker, F.A.:* J. Amer. Chem. Soc. *92*, 4235 (1970).
144. *Ochiai, E.I.:* J. Inorg. Nucl. Chem. *35*, 1727 (1973).
145. *Abel, E.W., Pratt, J.M., Whelan, R.:* J.C.S. Chem. Commun. *1971*, 449.
146. *Dedieu, A., Veillard, A.:* Theoret. Chim. Acta *36*, 231 (1975).
147. *Cockle, S.A., Hill, H.A.O., Williams, R.J.P.:* Inorg. Nucl. Chem. Letts. *6*, 131 (1970).
148. *Melamud, E., Silver, B.L., Dori, Z.:* J. Amer. Chem. Soc. *96*, 4689 (1974).
149. *Vansant, E.F., Lunsford, J.H.:* Adv. Chem. Ser. *No. 121*, 441 (1973).

150. *Tovrog, B. S., Drago, R. S.:* J. Amer. Chem. Soc. *96*, 6765 (1974).
151. *Hoffman, B. M., Szymanski, T., Basolo, F.:* J. Amer. Chem. Soc. *97*, 673 (1975).
152. *Getz, D., Melamud, M., Silver, B. L., Dori, Z.:* J. Amer. Chem. Soc. *97*, 3846 (1975).
153. *Calligaris, M., Nardin, G.:* Inorg. Nucl. Chem. Letts. *9*, 419 (1973).
154. *Rodley, G. A., Robinson, W. T.:* Nature *235*, 438 (1972).
155. *Collman, J. P., Takaya, H., Winkler, B., Libit, L., Koon, S. S., Rodley, G. A., Robinson, W. T.:* J. Amer. Chem. Soc. *95*, 1656 (1973).
156. *Brown, L. D., Raymond, K. N.:* Inorg. Chem. *14*, 2595 (1975).
157. *Brown, L. D., Raymond, K. N.:* J.C.S. Chem. Commun. *1974*, 470.
158. *Costa, G., Puxeddu, A., Nardin Stefani, L.:* Inorg. Nucl. Chem. Letts. *6*, 191 (1970).
159. *Carter, M. J., Engelhardt, L. M., Rillema, D. P., Basolo, F.:* J. C. S. Chem. Commun. *1973*, 810.
160. *Carter, M. J., Rillema, D. P., Basolo, F.:* J. Amer. Chem. Soc. *96*, 392 (1974).
161. *Tauzher, G., Amiconi, G., Antoninini, E., Brunori, M., Costa, G.:* Nature *241*, 222 (1973).
162. *Abel, E. W., Pratt, J. M., Whelan, R.:* Inorg. Nucl. Chem. Letts. 7 901 (1971).
163. *Swinehart, J. H.:* J. C. S. Chem. Commun. *1971*, 1443.
164. *Vaska, L.:* Science *140*, 809 (1963).
165. *Vaska, L., Rhodes, R. E.:* J. Amer. Chem. Soc. *87*, 4970 (1965).
166. *Flynn, B. R., Vaska, L.:* J. C. S. Chem. Commun. *1974*, 703.
167. *Parshall, G. W., Jones, F. N.:* J. Amer. Chem. Soc. *87*, 5356 (1965).
168. *Chock, P. B., Halpern, J.:* J. Amer. Chem. Soc. *88*, 3511 (1966).
169. *Ibers, J. A., La Placa, S. J.:* J. Amer. Chem. Soc. *87*, 2581, (1965).
170. *McGinnety, J. A., Doedens, R. J., Ibers, J. A.:* Science *155*, 709 (1967).
171. *McGinnety, J. A., Ibers, J. A.:* J. C. S. Chem. Commun. *1968*, 235.
172. *Terry, N. W., Amma, E. L., Vaska, L.:* J. Amer. Chem. Soc. *94*, 653 (1972).
173. *Weininger, M. S., Taylor, I. F., Amma, E. L.:* J. C. S. Chem. Commun. *1971*, 1172.
174. *Vaska, L., Chen, L. S.:* J. C. S. Chem. Commun. *1971*, 1080.
175. *Vaska, L., Chen, L. S., Senoff, C. V.:* Science *174*, 587 (1971).
176. *Peterson, W. M.:* Diss. Abs. Int. *B 34 (4)*, 1417 (1973).
177. *McGinnety, J. A., Payne, N. C., Ibers, J. A.:* J. Amer. Chem. Soc. *91*, 6301 (1969).
178. *Cook, C. D., Wan, K. Y., Gelius, U., Hamrin, K., Johansson, G., Olsson, E., Siegbahn, H., Nordiing, C., Siegbahn, K.:* J. Amer. Chem. Soc. *93*, 1904 (1971).
179. *Nolte, M. J., Singleton, E., Laing, M.:* J. Amer. Chem. Soc. *97*, 6396 (1975).
180. *Klevan, L., Peone, Jun., J., Madan, S. K.:* J. Chem. Ed. *50*, 670 (1973).
181. *Vaska, L., Chen, L. S., Miller, W. V.:* J. Amer. Chem. Soc. *93*, 6671 (1971).
182. *Hall, M. C., Kilbourn, B. T., Taylor, K. A.:* J. Chem. Soc. (A) *1970*, 2539.
183. *Kashiwagi, T., Yasuoka, N., Kasai, N., Kakudo, M., Takahashi, S., Hagihara, N.:* J. C. S. Chem. Commun. *1969*, 743.
184. *Takao, K., Fujiwara, Y., Imanaka, T., Yamamoto, M., Hirota, K., Teranishi, S.:* Bull. Chem. Soc. Japan *43*, 2249 (1970).
185. *Otsuka, S., Nakamura, A., Tatsuno, Y.:* J. Amer. Chem. Soc. *91*, 6994 (1969).
186. *Hirota, K., Yamamoto, M., Otsuka, S., Nakamura, A., Tatsuno, Y.:* J. C. S. Chem. Commun. *1968*, 533.
187. *Nakanura, A., Tatsuno, Y., Yamamoto, M., Otsuka, S.:* J. Amer. Chem. Soc. *93*, 6052 (1971).
188. *Horn, R. W., Weissberger, E., Collman, J. P.:* Inorg. Chem. *9*, 2367 (1970).
189. *Huber, H., Klotzbücher, W.:* Canad. J. Chem. *51*, 2722 (1973).
190. *Ozin, G. A., Klotzbücher, W. E.:* J. Amer. Chem. Soc. *97*, 3965 (1975).
191. *Klotzbücher, W., Ozin, G. A.:* J. Amer. Chem. Soc. *95*, 3790 (1973).
192. *Alexandrov, Y. A.:* J. Organometallic Chem. *57*, 71 (1973).
193. *Halpern, J., Goodall, B. L., Khare, G. P., Lim, H. S., Pluth, J. J.:* J. Amer. Chem. Soc. *97*, 2301 (1975).
194. *Wilke, G., Scott, H., Heinback, P.:* Angew. Chem. Int. Ed. Engl. *6*, 92 (1967).
195. *Cotton, F. A., Wilkinson, G.:* Advanced Inorganic Chemistry, 3rd Ed., New York: Wiley-Interscience 1972.

196. *Perutz, M.F., Muirhead, H., Cox, J.M., Goaman, L.G.:* Nature *219*, 131 (1968).
197. *Kendrew, J.C., Watson, H.C., Strandberg, B.E., Dickerson, R.E., Phillips, D.C., Shore, V.C.:* Nature *185*, 422 (1960).
198. *Perutz, M.F.:* Nature *228*, 726 (1970).
199. *Ibers, J.A., Lauher, J.W., Little, R.G.:* Acta Cryst. *B 30*, 268 (1974).
200. *Perutz, M.F.:* Nature *228*, 734 (1970).
201. *Deal, W.J.:* Biopolymers *12*, 2057 (1973).
202. *Hoard, J.L.:* Structural Chemistry and Molecular Biology (edit. Rich. A., and Davidson, N.). San Francisco: Freeman 1968.
203. *Corwin, A.H., Reyes, Z.:* J. Amer. Chem. Soc. *78*, 2437 (1956).
204. *Wang, J.H.:* J. Amer. Chem. Soc. *80*, 3168 (1958).
205. *Mason, R.:* Nature *217*, 543 (1968).
206. *Mingos, D.M.P.:* Nature *229*, 194 (1971).
207. Ibid. *230*, 154 (1971).
208. *Hodgson, D.J., Payne, N.C., McGinnety, J.A., Pearson, R.G., Ibers, J.A.:* J. Amer. Chem. Soc. *90*, 4486 (1969).
209. *Snyder, D.A., Weaver, D.L.:* J.C.S. Chem. Commun. *1969*, 1425.
210. *Mingos, D.M.P., Ibers, J.A.:* Inorg. Chem. *10*, 1035, 1043 (1971).
211. Ibid., p. 1479.
212. *Erskine, R.W., Field, B.O.:* Current Investigations.
213. *Weiss, J.J.:* Nature *202*, 83 (1964).
214. *Viale, R.O., Maggiora, G.M., Ingraham, L.L.:* Nature *203*, 183 (1964).
215. *Halton, M.P.:* Theoret. Chim. Acta (Berl.) *24*, 89 (1972).
216. *Halton, M.P.:* Inorganica Chim. Acta *8*, 131 (1974).
217. *Cockle, S.A., Hill, H.A.O., Pratt, J.M., Williams, R.J.P.:* Biochem. Biophys. Acta *177*, 686 (1969).
218. *Halton, M.P.:* Inorganica Chim. Acta *8*, 137 (1974).
219. *Trautwein, A.:* In: Perspectives in Mössbauer Spectroscopy (eds. S.C. Cohen and M. Pasternak). New York: Plenum Press 1973.
220. *Zerner, M., Gouterman, M., Kobayashi, H.:* Theor. Chim. Acta *6*, 363 (1966).
221. *Weissbluth, M., Maling, J.E.:* J. Chem. Phys. *47*, 4166 (1967).
222. *Trautwein, A.:* Struct. Bonding *20*, 101 (1974).
223. *Tasaki, A., Otsuka, J.:* IEEE Trans. Magn. *9*, 389 (1973).
224. *Otsuka, J., Matsuoka, O., Fuchikami, N., Seno, Y.:* J. Phys. Soc. Japan *35*, 854 (1973). – Ibid. *33*, 1645 (1972).
225. *Weissbluth, M.:* Struct. Bonding *2*, 1 (1967).
226. *Rossi-Fanelli, A., Antonini, E., Caputo, A.:* Advan. Protein Chem. *19*, 73 (1964).
227. *Lang, G., Marshall, W.:* Proc. Phys. Soc. *87*, 3 (1966).
228. *Chantrell, S.J., McAuliffe, C.A., Munn, R.W., Pratt, A.C.:* Co-ordination Chem. Rev. *16*, 259 (1975).
229. *Zerner, M., Gouterman, M.:* Theor. Chim. Acta *4*, 44 (1966).
230. *Gonzalez, B., Kouba, J., Yee, S., Reed, C.A., Kirner, J.F., Scheidt, W.R.:* J. Amer. Chem. Soc. *97*, 3247 (1975).
231. *Barlow, C.H., Maxwell, J.C., Wallace, J.W., Gaughey, W.S.:* Biochem. Biophys. Res. Comm. *55*, 91 (1973).
232. *Collman, J.P., Gagne, R.R., Reed, C.A., Halbert, T.R., Lang, G., Robinson, W.T.:* J. Amer. Chem. Soc. *97*, 1427 (1975).
233. *Collman, J.P., Reed, C.A.:* J. Amer. Chem. Soc. *95*, 2048 (1973).
234. *Loew, G.H., Kirchner, R.F.:* J. Amer. Chem. Soc. *97*, 7388 (1975).
235. *Collman, J.P., Gagne, R.R., Halbert, T.R., Marchon, J.C., Reed, C.:* J. Amer. Chem. Soc. *95*, 7868 (1973).
236. *Almog, J., Baldwin, J.E., Dyer, R.L., Peters, M.:* J. Amer. Chem. Soc. *97*, 226 (1975).
237. *Collman, J.P., Gagne, R.R., Reed, C.A., Robinson, W.T.:* Proc. Nat. Acad. Sci. U.S.A. *71*, 1326 (1974).

238. *Gottwald, L.K., Ullman, E.F.:* Tetrahedron Letts. *1969*, 3071.
239. *Lindoy, L.F.:* Chem. Soc. Rev. *4*, 421 (1975).
240. *Collman, J.P., Gagne, R.R., Gray, H.B., Hare, J.W.:* J. Amer. Chem. Soc. *96*, 6522 (1974).
241. *Collman, J.P., Gagne, R.R., Reed, C.A.:* J. Amer. Chem. Soc. *96*, 2629 (1974).
242. *Chang, C.K., Traylor, T.G.:* J. Amer. Chem. Soc. *95*, 5810 (1973).
243. *Epstein, L.M., Straub, D.K., Maricondi, C.:* Inorg. Chem. *6*, 1720 (1967).
244. *Goedken, V.L.:* J.C.S. Chem. Commun. *1972*, 207.
245. *Eicher, H., Trautwein, A.:* J. Chem. Phys. *50*, 2540 (1969).
246. *Anderson, D.L., Weshler, C.J., Basolo, F.:* J. Amer. Chem. Soc. *96*, 5599 (1974).
247. *Hoffman, A.B., Collins, D.M., Day, V.W., Fleisher, E.M., Srivastava, T.S., Hoard, J.L.:* J. Amer. Chem. Soc. *94*, 3620 (1972).
248. *Fleisher, E.B., Srivastava, T.S.:* J. Amer. Chem. Soc. *91*, 2403 (1969).
249. *Alben, J.O., Fuchsman, W.H., Beaudreau, C.A., Caughey, W.S.:* Biochemistry 7, 624 (1968).
250. *Kao, O.H.W., Wang, J.H.:* Biochemistry *4*, 342 (1965).
251. *Weschler, C.J., Anderson, D.L., Basolo, F.:* J. Amer. Chem. Soc. *97*, 6707 (1975).
252. *Stynes, H.C., Ibers, J.A.:* J. Amer. Chem. Soc. *94*, 5125 (1972).
253. *Burwell, Jun., R.L.:* Chem. Technol. *1974*, 370.
254. *Leal, O., Anderson, D.L., Bowman, R.G., Basolo, F., Burwell, R.L.:* J. Amer. Chem. Soc. *97*, 5125 (1975).
255. *Nobbs, C.L., Watson, H.C., Kendrew, J.C.:* Nature *209*, 339 (1966). – *Kendrew, J.C., Watson, H.C., Strandberg, B.E., Dickerson, R.E., Phillips, D.C., Shore, V.C.:* Ibid. *190*, 666 (1961). – *Watson, H.C.:* Progr. Stereochem. *4*, (1968).
256. *Bull, C., Fisher, R.G., Hoffman, B.M.:* Biochem. Biophys. Res. Comm. *59*, 140 (1974).
257. *Weschler, C.J., Hoffman, B.M., Basolo, F.:* J. Amer. Chem. Soc. *97*, 5278 (1975).
258. *Wayland, B.B., Minkiewicz, J.V., Abd-Elmageed, M.E.:* J. Amer. Chem. Soc. *96*, 2795 (1974).
259. *Wagner, G.C., Kassner, R.J.:* J. Amer. Chem. Soc. *96*, 5593 (1974).
260. *Almog, J., Baldwin, J.E., Huff, J.:* J. Amer. Chem. Soc. *97*, 227 (1975).
261. *Almog, J., Baldwin, J.E., Dyer, R.L., Huff, J., Wilkerson, C.J.:* J. Amer. Chem. Soc. *96*, 5560 (1974).
262. *Chang, C.K., Traylor, T.G.:* Proc. Nat. Acad. Sci. U.S.A. *70*, 2647 (1973).
263. *Chang, C.K., Traylor, T.G.:* J. Amer. Chem. Soc. *95*, 8475 (1973).
264. *Brinigar, W.S., Chang, C.K., Geibel, J., Traylor, J.G.:* J. Amer. Chem. Soc. *96*, 5597 (1974).
265. *Chang, C.K., Traylor, T.G.:* J. Amer. Chem. Soc. *95*, 8477 (1973).
266. *Brinigar, W.S., Chang, C.K.:* J. Amer. Chem. Soc. *96*, 5595 (1974).
267. *Hsu, G.C., Spilburg, C.A., Bull, C., Hoffman, B.F.:* Proc. Nat. Acad. Sci. U.S.A. *69*, 2122 (1972).
268. *Spilburg, C.A., Hoffman, B.M., Petering, D.H.:* J. Biol. Chem. *247*, 4219 (1972).
269. *Walker, F.A.:* J. Amer. Chem. Soc. *95*, 1154 (1973).
270. *Hill, H.A.O.:* J. Organometallic Chem. *11*, 167 (1968).
271. *Hoffman, B.M., Ptering, D.H.:* Proc. Nat. Acad. Sci. U.S.A. *67*, 637 (1960).
272. *Stynes, H.C., Ibers, J.A.:* J. Amer. Chem. Soc. *94*, 1559 (1972).
273. *Stynes, D.V., Stynes, H.C., Ibers, J.A., James, B.R.:* J. Amer. Chem. Soc. *95*, 1142 (1973).
274. *Stynes, D.V., Stynes, H.C., James, B.R., Ibers, J.A.:* J. Amer. Chem. Soc. *95*, 1796 (1973).
275. *Gindry, R.M., Drago, R.S.:* J. Amer. Chem. Soc. *95*, 6645 (1973).
276. *Ibers, J.A., Stynes, D.V., Stynes, H.C., James, B.R.:* J. Amer. Chem. Soc. *96*, 1358 (1974).
277. *Beugelsdijk, T.J., Drago, R.S.:* J. Amer. Chem. Soc. *97*, 6466 (1975).
278. *Collman, J.P., Sorrell, T.N., Hoffman, B.M.:* J. Amer. Chem. Soc. *97*, 913 (1975).
279. *Koch, S., Tang, S.C., Holm, R.H., Frankel, R.B., Ibers, J.A.:* J. Amer. Chem. Soc. *97*, 916 (1975).
280. *Chang, C.K., Dolphin, D.:* J. Amer. Chem. Soc. *97*, 5948 (1975).
281. *Geibel, J., Chang, C.K., Traylor, T.G.:* J. Amer. Chem. Soc. *97*, 5924 (1975).
282. *Collman, J.P., Brauman, J.I., Suslick, K.S.:* J. Amer. Chem. Soc. *97*, 7185 (1975).

Structural and Bonding Aspects in Phosphorus Chemistry-Inorganic Derivatives of Oxohalogeno Phosphoric Acids

Kurt Dehnicke and Abdel-Fattah Shihada

Fachbereich Chemie der Universität Marburg, D-3550 Marburg/Lahn, Lahnberge

Table of Contents

1. Introduction

The halophosphoric acids and -phosphates of the type $[PO_3F]^{2\ominus}$, $[PO_2F_2]^{\ominus}$, $[POSF_2]^{\ominus}$, $[PO_2Cl_2]^{\ominus}$, $[POSCl_2]^{\ominus}$, $[PO_2Br_2]^{\ominus}$ and $[PO_2FY]^{\ominus}$ (Y = Cl, Br), which are to be treated here, are formally derived from the $[PO_4]^{3\ominus}$-ion by substitutuion of one or several O-atoms by halogen or sulfur. But whereas in the orthophosphate ion coordination to other elements is only possible by the four O-atoms, which cannot be distinguished, the halophosphates show different coordination behaviour according to the different substituents. Actually the O-atoms coordinate preferably in the halophosphates too, but in case of the fluorophosphates there is some competition between O- and F-atoms in coordination behaviour, which seems reasonable because of their compasable electronegativity.

The smaller electrostatic charge of the halophosphates, i.e. $2\ominus$ and $1\ominus$ resp., compared with the triple charge in $[PO_4]^{3\ominus}$, gives rise to a certain anisotropy of the bonding system, which especially in the dihalophosphates causes the formation of bonds with evidently homopolar character. This is certainly one of the reasons for their structural variety.

Some attention is to be paid to the isoelectronic species too, the series $PO_4^{3\ominus}$, $PO_3F^{2\ominus}$, $PO_2F_2^{\ominus}$, POF_3, $SO_4^{2\ominus}$, $SO_3F^{\ominus}$, SO_2F_2 for example allows interesting comparisons as far as structure and bonding is concerned.

2. Free acids

The latest review about the free haloacids and their formation has been given by *Grunze* (*1*) in 1963 and an extensive treatment of the fluoro derivatives is found in the work of *Schmutzler* (*2*). The only halophosphoric acids prepared up to now are HPO_2F_2, H_2PO_3F, HPO_2Cl_2 and the mixed fluoro haloacids HPO_2FX (X = Cl, Br), the most important properties of which are listed in Table 1.

Table 1. Halophosphoric acids known in substance

	Melting point [°C]	Boiling point [°C]	Thermal stability [°C]	d_4^{20}	Characteristics	Ref.
HPO_2F_2	−91,8 to −93,1	116,5[a])	up to 100	1,63	colourless, viscous, degree of association in the gas phase 2,7	(*3*), (*4*), (*5*)
H_2PO_3F	–	–	up to 185	1,90	colourless, viscous	(*3*), (*5*)
HPO_2Cl_2	−28	–	stable in the liquid phase at room temp.	1,686 (25°)	colourless, fluid liquid, dimeric	(*6*)
HPO_2FCl	–	215[a])	Distillation in vacuo and	1,522	colourless, fluid liquids	(*7*)
HPO_2FBr	–	–	below 50 °C possible	2,140		(*7*)

[a]) Extrapolated from measurements of vapour pressure

HPO_2Br_2 (*8*), $HPOSF_2$ (*9*) and $HPOSCl_2$ (*10*) have been prepared in solution exclusively, so that a characterisation was possible only indirectly by NMR and absorption spectroscopy or by preparation of more stable derivates, such as insoluble salts of these acids. Here it might be suitable to mention the likewise unstable peroxo halophosphoric acids H_2PO_4F and HPO_3F_2, which have been prepared by *Fluck et al.* (*11*) by reaction of H_2O_2 with $POCl_2F$ and $P_2O_3F_4$ resp.

Considering the different ways of formation of the halophosphoric acids, it should be stated that the partial hydrolysis of the corresponding oxo- and thiohalogen compounds resp. yields the free acids according to reaction (1), but the products are of considerable impurity because of the uncontrollability of the process

$$OPX_3 + H_2O \rightarrow OPX_2OH + HX \qquad (1)$$

and because of secondary reactions of the hydrogen halide. So actually the following methods are considered most effective for preparation:

Difluorophosphoric acid has first been prepared by *Lange et al.* (*3*) and is now being prepared commercially by the reaction:

$$P_4O_{10} + 6\,HF \rightarrow 2\,HPO_2F_2 + 2\,H_2PO_3F \tag{2}$$

The difluorophosphoric acid can be separated from the nonvolatile monofluoro compound by vacuum distillation. The product is purified over P_4O_{10} and by condensing at – 78 °C under high vacuum (*4*).

In the absence of moisture difluorophosphoric acid is stable for a long time; it does not decompose until being heated to about 100 °C:

$$2\,HPO_2F_2 \rightarrow POF_3 + H_2PO_3F \tag{3}$$

I.r. spectra (*4, 13*) and measurements of vapor pressure (*4*) indicate strong intermolecular hydrogen bridging. In contrast to concentrated H_2SO_4 difluorophosphoric acid is a base and can be titrated conductometrically (*14*):

$$HPO_2F_2 + H_2SO_4 \rightleftarrows H_2PO_2F_2^{\oplus} + HSO_4^{\ominus} \tag{4}$$

Monofluorophosphoric acid is available by treatment of metaphosphoric acid with anhydrous HF (*15*):

$$HPO_3 + HF \rightarrow H_2PO_3F \tag{5}$$

In many properties the colour- and nonodorous nonvolatile acid, which is highly viscous, resembles remarkably the isoelectronic H_2SO_4. Thus the densities are very similar and further on $CdSO_4 \cdot 8/3\,H_2O$ and $CdPO_3F \cdot 8/3\,H_2O$ (*16*) are isostructural. Also, H_2PO_3F is relatively stable kinetically, and can be isolated from aqueous solutions of its salts by ion exchange, which is an adequate method e.g. for preparation of the acid salts M^IHPO_3 (*17*). The high viscosity as well as the nonvolatility suggest strong intermolecular hydrogen bridging.

Dichlorophosphoric acid is most conveniently prepared by the careful hydrolysis of the easily accessible (*18*) anhydride at – 30 °C (*6*):

$$P_2O_3Cl_4 + H_2O \rightarrow 2\,HOPOCl_2 \tag{6}$$

An extremely pure product results, when difluorochloro methane or difluorodichloro methane are used as solvents (*19*). Dichlorophosphoric acid is a fluid, colourless, very hygroscopic liquid, which is easily soluble in $CHCl_3$, CCl_4, Ethanol and Ether (*6*). In the liquid phase it is stable for some time at room temperature, whereas at 12 Torr there is no sign of decomposition up to 250 °C (*6*). According to the Raman spectra in the liquid it is dimeric in analogy to the carboxylic acids (*20*):

$$\begin{array}{c} Cl \\ Cl \end{array} > P \begin{array}{l} {=}O{---}H{-}O \\ {-}O{-}H{---}O{=} \end{array} P < \begin{array}{c} Cl \\ Cl \end{array}$$

The different behaviour of difluorophosphoric acid which is of a higher degree of association, is probably induced by the ability of the fluorine ligands to act, besides the O-atoms, as bridging atoms in the hydrogen bridges.

A chance to produce certain dihalophosphoric acids with two different halogen atoms, showed up, when *Falius* (*21*) succeeded in preparing "monofluorophosphorous acid" by hydrolysis of PCl_3 in presence of hydrogen fluoride:

$$PCl_3 + 2\,H_2O + HF \xrightarrow{0^\circ} H[HPO_2F] + 3\,HCl \qquad (7)$$

In cold etheric solution it undergoes redox-reactions with Cl_2 and Br_2 dissolved in CCl_4, such as (8) (*7*):

$$H[HPO_2F] + X_2 \rightarrow H[PO_2FX] + HX \qquad (8)$$

After evaporating the solvents and hydrogen halides the free acids can be purified by fractional distillation in vacuo. It was not possible to isolate the acid with X = J (*7*). The high value of the Trouton constant (25.1 cal/Mol · °K) (*7*) seems to indicate, that $H[PO_2FCl]$ and $H[PO_2FBr]$ as well are associated over hydrogen bridges.

At present there is no more than indirect evidence for the existence of the *halogeno-halothiophosphoric acids* and the *dibromo compound*. So the unsoluble nitron dibromophosphate precipitates in case of the partial hydrolysis of $POBr_3$ in acetone and in presence of nitron (*8*). Further the partial hydrolysis of $PSCl_3$ and $PSCl_2F$ in presence of big cations, such as tetraphenylphosphonium or -arsonium yields stable salts (*22*), which suggests the primary formation of the free acid in the solution:

$$PSCl_3 + H_2O + [(C_6H_5)_4P]Cl \rightarrow [(C_6H_5)_4P]^{\oplus}[OSPCl_2]^{\ominus} + 2\,HCl \qquad (9)$$

$$PSCl_2F + H_2O + [(C_6H_5)_4As]Cl \rightarrow [(C_6H_5)_4As]^{\oplus}[OSPClF]^{\ominus} + 2\,HCl \qquad (10)$$

3. Inorganic derivatives

3.1. Monofluorophosphates $[PO_3F]^{2\ominus}$

The monofluorophosphates, the sodium salt of which is used in caries prophylaxis (*23*), are very similar to the isoelectronic sulfates in their macroscopic properties (*24*). This becomes evident looking at the solubilities of these compounds: the fluorophosphates $M_2^I[PO_3F]$ mit M^I = alkaly and ammonium are easily soluble in water and the solubility of the Ba-, Pb- etc. salts is equally comparable to that of the corresponding sulfates. Similarly the composition of the hydrates corresponds very well to stochiometry of the sulfates as far as crystal water is concerned and can be demonstrates by the examples: $MnPO_3F \cdot 4\,H_2O$, $NiPO_3F \cdot 7\,H_2O$, $Cr_2(PO_3F)_3 \cdot 18\,H_2O$ (*25*). Induced these compounds are isostructural with the corresponding sulfates. Structural investigations however reveal significant differences between sulfate and fluorophosphate also, which are induced by the nature of the P–F bond, which, in contrast to the P–O bond, does not contain significant $(d_\pi - p_\pi)$-contributions. As a consequence this kind of bond strengthening holds essentially for the three O-ligands of the $[PO_3F]^{2\ominus}$ ion only:

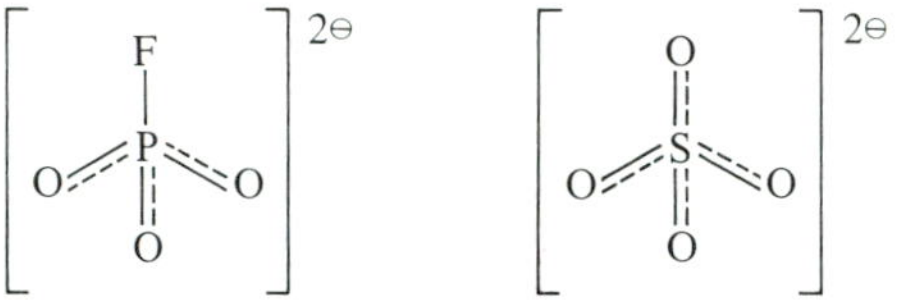

The different P–O and P–F bond lengths cause a quite complex coordination behaviour. So the average P–O bond length in $CaPO_3F \cdot 2\,H_2O$ ist 1.508 Å, whereas rP–F amounts to 1.583 Å (*26*); in contrast the average bond length in $CaSO_4 \cdot 2\,H_2O$ (gypsum) is 1.489 Å (*27*) however. As the F-atom in the fluorophosphate in $CaPO_3F \cdot 2\,H_2O$ neither participates in Hydrogen bridges nor is coordinated to the $Ca^{2\oplus}$ ion, the result is a seven-coordination for calcium (5 O-atoms from four different PO_3F-groups and two O-atoms from two H_2O molecules) (*26*), while a coordination number of 8 is found for calcium in $CaSO_4 \cdot 2\,H_2O$ (6 O-atoms from 4 sulfate ions + 2 H_2O) (*27*).

The different coordination behaviour of F- and O-ligands in the $[PO_3F]^{2\ominus}$ ion becomes evident in the interesting structural isomery of $(NH_4)_2PO_3F \cdot H_2O$ (*26*) and $(NH_4)_2SO_3 \cdot H_2O$ (*28*) also, the structural parameters of which are very much alike – the part of the F-atom in the fluorophosphate being played by the lone electron pair in the sulfite ion.

For the *preparation of monofluorophosphates* there are quite a lot of methods found in literature. The alkali salts are obtained by shortly melting poly- or trimetaphosphate with alkali fluoride (*29, 30*) – the application of graphite vessels yielding particulary pure products (*31*):

$$(NaPO_3)_n + n\,NaF \rightarrow n\,Na_2[PO_3F] \qquad (11)$$

The unsoluble fluorophosphates can be prepared from the sodium salt by reaction with heavy metal nitrates, whereas the watersoluble salts are appropriately prepared by the reaction of the sligthly soluble Ag_2PO_3F with heavy metal chlorides (*24*). Acid salts $M^IH[PO_3F]$ with M^I = Na, K, NH_4 result from fractional precipitation by addition of alcohol and ether to the aqueous solution of the free acid (*17*). Further methods for the synthesis of fluorophosphates are given by Eqs. (12) and (13), which lead to dimethylin fluorophosphate (*32*) and to tin (II) difluorophosphate (*33*):

$$(CH_3)_2SnCl_2 + K_2PO_3F \rightarrow (CH_3)_2SnPO_3F + 2\,KCl \qquad (12)$$

$$SnF_2 + H_2PO_3F \rightarrow SnPO_3F + 2\,HF \qquad (13)$$

3.2. Difluorophosphates $[PO_2F_2]^\ominus$

In some of its properties the difluorophosphate ion resembles the isoeletronic perchlorate ion, although the potassium and the caesium salt are easily soluble in water (*34*). However there are significant differences, especially in structure evident also, which can be attributed to the different lengths of the P–O and the P–F bonds. The bond length anisotropy is more distinct in $K[PO_2F_2]$ (r_{PO}: 1.40 Å, r_{PF}: 1.575 Å (*35*)) than in the $[PO_3F]^{2\ominus}$ ion. Undoubtedly this is a consequence of the concentration of π-bonding on but two P–O bonds, which is an explanation for the different bond angle OPO (122.4°) and FPF (97.1°) as well:

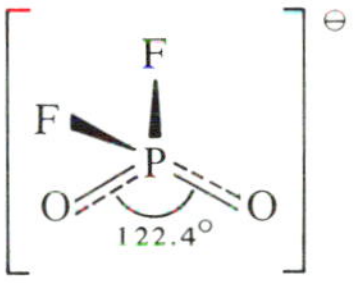

122.4°

(I)

As a matter of fact the interest in the difluorophosphate group partially results from its structural variety. So beside ion (I) the monofunctional $OP(O)F_2$ group (II), which includes the anhydride of the difluorophosphoric acid $P_2O_3F_4$ (*36*), and the O_2PF_2 group (III) acting as a bridge between two metal atoms, are known:

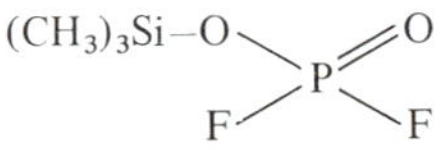

$(CH_3)_3Si{-}O{-}P(O)F_2$ $[(CH_3)_2Ga(O_2PF_2)]_2$

(II) [37), 38)] 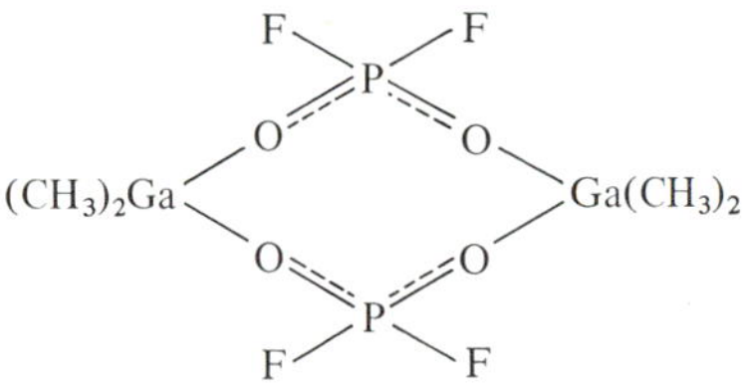   (III) [39)]

The type of bonding in (III) may give rise to the formation of polymers, realised by $Si(O_2F_2P)_4$ (*38*) and $Fe(O_2PF_2)_3$ (*40, 41*) for example. In the ^{57}Fe-Mössbauer spectra of the iron compound an unusually broadened line and a large quadrupol splitting compared to $Fe(PO_2Cl_2)_3$ indicate the coordinative participation of the fluorine atoms in the octahedral environment of the iron atoms (*41*). Yet the $[PO_2F_2]^{\ominus}$ ion does not show any tendency towards the formation of chelates, which are well known from the $[PO_2Cl_2]^{\ominus}$ ion. The reason might be found in the increased OPO bond angle (122.4°), which would cause too much stress towards forming a four-ring system. On the other hand the trialkyltin-difluorophosphates $(CH_3)_3Sn[O_2PF_2]$ (*42*) and $(C_4H_9)_3Sn[O_2PF_2]$ (*38*) are evidently monomeric molecules, in which a coordination number 4 for the tin atoms, i.e. a bond type (II) for the difluorophosphate group, should be quite improbable. So it is possible that, besides an O-atom, a F-atom acts as coordinating ligand in these compounds, because the OPF bond angle of 108° (*35*) is geometrically more favourable than the larger OPO bond angle.

In agreement with this bonding concept is the structure of the polymeric diakyltin(bis)difluorophosphates $R_2Sn(O_2PF_2)_2$ with R = CH_3, C_2H_5, n–C_3H_7 etc., in which, according to the vibrational and the ^{119}Sn-Mössbauer spectra, the tin atoms are octahedrally coordinated with the O_2PF_2 groups acting as bridging ligands of the type (III) (*43*).

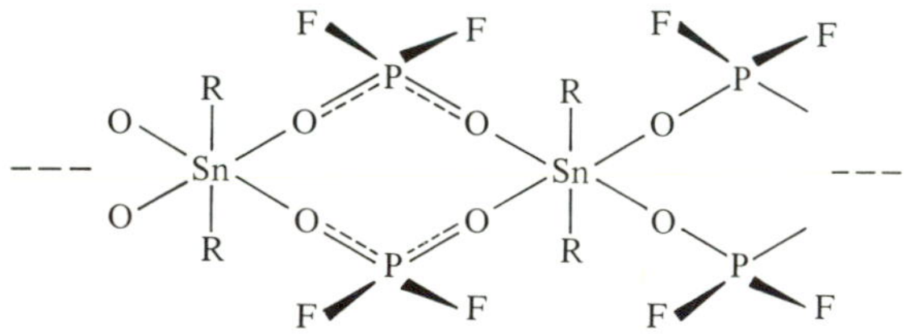

Naturally there are different factors which determine the one of the three structure types of the difluorophosphate group, which actually occurs. However, a rough classification is possible, according to which cations with relatively low ion potential (e.g. alkali ions) induce the structure type (I), whereas the presence of groups causing a strong anisotropy at the cation site, such as R_3Si (*37*) or F–Xe (*44*) usually lead to the structure type (II); finally for the formation of type (III), additional coordinative possibilities of the element bonded to the PO_2F_2 group are necessary.

The determination of the actually occuring structure types is can be done by vibrational spectroscopy in most cases (s. section 4.2.).

The *difluorophosphates are prepared* by various methods. Very widely used is the reaction of difluorophosphoric acid with metal chlorides according to eqn. (14):

$$MCl + HPO_2F_2 \rightarrow M[O_2PF_2] + HCl \qquad (14)$$

The volatility of the hydrogen chloride allows the quantitative reaction of the components. In this way the following difluorophosphates were prepared:

$M^I[PO_2F_2]$ (M^I = Li, Na, K, Rb, Cs, NH_4) (*45*); $NO[PO_2F_2]$ (*46*);
$Ca[O_2PF_2]_2 \cdot 2$ Acetic acid ethyl ester (*47*); $Cl_2Al[PO_2F_2]$ (*48*);
$R_2Sn[PO_2F_2]_2$ (R = CH_3, C_2H_5, n–C_3H_7, n–C_4H_9, n–C_8H_{17}) (*43*);
$(CH_3)_3Si[O_2PF_2]$ (*49*); $Cl_2Ti[PO_2F_2]_2$ (*50*).

The dimethyl difluorophosphates $(CH_3)_2M[PO_2F_2]$ with M = Al, Ga, In and Tl (*39*) are readily formed by reaction (15):

$$(CH_3)_3M + HPO_2F_2 \rightarrow (CH_3)_2M[PO_2F_2] + CH_4 \quad (15)$$

In some cases the use of ammonium difluorophosphate is suitable. The reaction is then performed in ethereal solution, from which the unsoluble ammonium chloride can be separated by filtration:

$$R_3SiCl + NH_4[PO_2F_2] \rightarrow R_3Si[PO_2F_2] + NH_4Cl \quad (16)$$

In this way difluorophosphates with mixed organic substituents R and R′ are obtained (*37*) as well as dialkylsilicon(bis)difluorophosphates $RR'Si[PO_2F_2]_2$ (*51*).

Furthermore reactions with difluorophosphoric anhydride are of some importance (*38*). The easy cleavage of the P–O–P-bridge in reactions with organometallic oxides (17) and halides (18) leads to derivatives of difluorophosphoric acid in a very mild way:

$$(CH_3)_3Si{-}O{-}Si(CH_3)_3 + P_2O_3F_4 \xrightarrow{+20^\circ} 2\,(CH_3)_3Si[PO_2F_2] \quad (17)$$

$$XeF_2 + P_2O_3F_4 \xrightarrow{-22^\circ} FXe[PO_2F_2] + POF_3 \quad (18)$$

By this scheme $(CH_3)_3Si[PO_2F_2]$ (*38*) and $FXe[PO_2F_2]$ (*44*), $(CH_3)_3Sn[PO_2F_2]$ (*42*), polymer $Si[PO_2F_2]_4$ (*38*) and $Xe[PO_2F_2]_2$ (*44*) are also accesible. On the other hand the cleavage of the Si–O–Si-bridge in hexamethyl disiloxane with POF_3 requires a temperature of about 100 °C (*49*):

$$(CH_3)_3Si{-}O{-}Si(CH_3)_3 + POF_3 \rightarrow (CH_3)_3Si[PO_2F_2] + (CH_3)_3SiF \quad (19)$$

It is interesting to note that even the Si–N–Si-bridge is cleaved with $P_2O_3F_4$ (*52*):

$$(CH_3)_3Si{-}NH{-}Si(CH_3)_3 + P_2O_3F_4 \rightarrow (CH_3)_3Si[PO_2F_2] + (CH_3)_3Si{-}NH{-}POF_2 \quad (20)$$

More complicated are fluorination reactions of phosphorus pentoxide with NH_4F (*34*) and $NaHF_2$ (*53*); yet they are suited for the preparation of fluorophosphates on large scale:

$$P_4O_{10} + 6\ NH_4F \rightarrow 2\ NH_4[PO_2F_2] + 2\ (NH_4)_2[PO_3F] \tag{21}$$

The ammonium difluorophosphate which is better soluble in ethanol is separated by extraction with this solvent (*34*). The reactions (22) (*54*) and (23) (*55*) which are run at higher temperatures do not give purer products:

$$KPF_6 + 2\ KPO_3 \xrightarrow{400^\circ} 3\ K[PO_2F_2] \tag{22}$$

$$2\ CsF + 2\ POF_3 \longrightarrow Cs[PO_2F_2] + CsPF_6 \tag{23}$$

Last, but not least, there is to be mentioned, that in some cases a direct transformation of dichlorophosphates (see section 3.4.) into difluorophosphates by fluorination with elementary fluorine ($In(PO_2F_2)_3$, $Fe(PO_2F_2)_3$ (*40*)) or with arsenic trifluoride at elevated temperature is possible (*41*):

$$Fe(PO_2Cl_2)_3 + 2\ AsF_3 \rightarrow Fe(PO_2F_2)_3 + 2\ AsCl_3 \tag{24}$$

3.3. Difluorothiophosphates $[POSF_2]^\ominus$

Considering ions of this type, the question arises, whether the P—S-bond participates in the resonance (I a) and (I b) with a comparable degree to the PO bond:

$$\left[\begin{matrix} O\!=\!P(-S)(F)_2 \end{matrix}\right]^\ominus \longleftrightarrow \left[\begin{matrix} O\!-\!P(=S)(F)_2 \end{matrix}\right]^\ominus$$

(Ia) (Ib)

For the ions with mixed halogen ligands $[POSFCl]^\ominus$ as well as $[POSCl_2]^\ominus$ with tetraphenyl phosphonium and -arsonium as cations it is assumed on account of infrared data, that "die Salzbildung am Schwefel erfolgt", (*22*)[1]) and hence type (I a) is existent. It is not possible however, to decide this problem by the aid of the frequencies of the PO and PS stretching vibrations, because νPS as well as νPO in the oxothiohalophosphates are shifted to longer wavelengths to about the same extent compared with $O{=}PX_3$ and $S{=}PX_3$ resp., where X = F, Cl (*56*). This holds in the case of the spectra of $[N(C_2H_5)_4]^\oplus[POSF_2]^\ominus$ too (*57*).

---

[1]) "the formation of the salt occurs at the sulfur atom"

In the structures of strongly anisotropic groups, as in the compounds $(CH_3)_3Si-O-P(S)F_2$ (*58*) and $(CH_3)_3SnOPSF_2$ (*42*), the O-atom is the favoured coordination ligand, whereas in the tin compound the sulfur atom coordinates too. Obviously sulfur coordination if favoured to F-coordination.

$(CH_3)_3Si-O-P(=S)F_2$ $\qquad$ $(CH_3)_3Sn(O,S)PF_2$

The structure of the following dithiomonofluorophosphate (*59*) corresponds to this rule:

$(CH_3)_3Sn(S,S)P(F)(C_2H_5)$

For the *preparation of halothiophosphates* the same methods as described for the difluorophosphates (see section 3.2.) can be used. The salts are obtained by partial hydrolysis of PSF_3 and PSF_2Cl resp. in presence of tetraphenyl phosphonium or -arsonium chloride according to eqns. (9, 10) (*22*), whereas $(CH_3)_3Si-O-P(S)F_2$ can be prepared corresponding to eqn. (17) by reaction of $F_2P(S)-O-P(S)F_2$ (*60*) with hexamethyl disiloxane at 80 °C (*58*). In the same way the reaction of $[(CH_3)_3Si]_2O$ with PSF_3 according to Eqn. (*19*) is for its preparation (*58*). For the preparation of $(CH_3)_3Sn[POSF_2]$ corresponding reaction of $[(CH_3)_3Sn]_2O$ with PSF_3 is found in literature (*42*). By an original synthesis tetraethyl ammonium thiodifluorophosphate had been obtained through reaction (25) already in 1965 (*57*):

$$F_2(S)P-O-C_2H_5 + N(C_2H_5)_3 \rightarrow [N(C_2H_5)_4]^{\oplus}[POSF_2]^{\ominus} \qquad (25)$$

3.4. *Dichlorothiophosphates* $[POSCl_2]^{\ominus}$

At present only the tetraphenylphosphonium and the -arsonium salt are described, which are accessible by the reaction of $PSCl_3$ with $[(C_6H_5)_4P]Cl$ and $[(C_6H_5)_4As]Cl$ resp. (*22*) corresponding to Eqs. (9, 10).

3.5. *Dichlorophosphates* $[PO_2Cl_2]^{\ominus}$

The most intensely investigated of all halophosphates are certainly the dichlorophosphates, about which extensive crystallographic and spectroscopic data exist. This might be brought into connection with the fact, that for the preparation of the dichlorophosphates there is an especially rich variety of methods for synthesis at hand. On the other hand the dichlorophosphate groups represent interesting objects of study as far as their different ways of coordination are concerned (I–V), going from the monofunctional dichlorophosphate group (I) to the ion (II) and the trifunctional structural element (V):

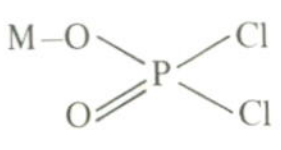

(I) 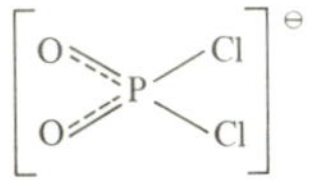 (II) 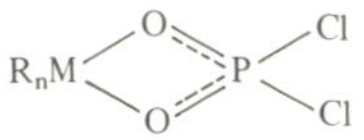 (III)

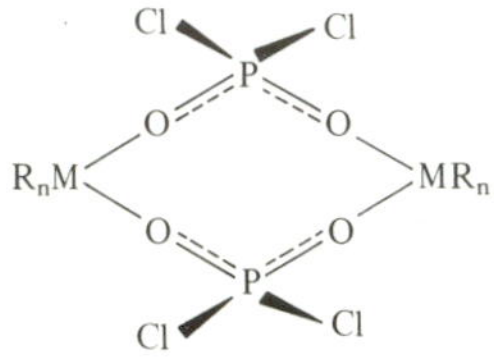

(IVa) 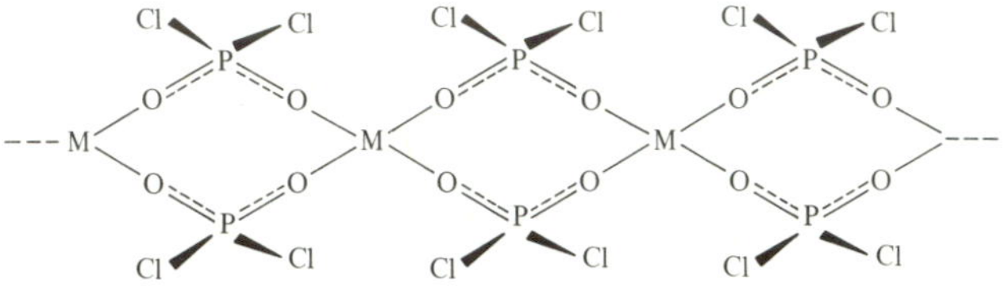 (IVb)

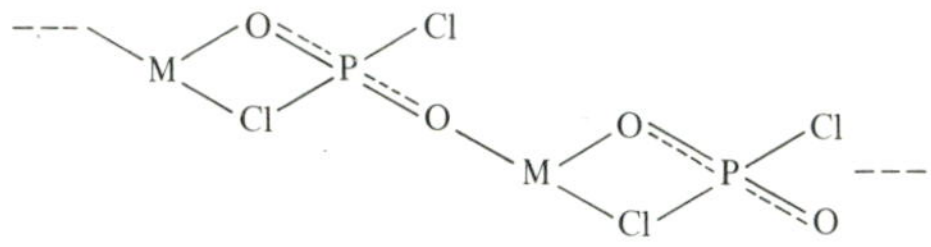

(V)

Table 2 gives a view of the inorganic dichlorophosphates, known at present, and the structural types of their dichlorophosphate groups.

The monofunctional type (I) is realised in molecules containing substituents with strongly anisotropic effects and without possibilities for an additional coordination, as for example in R_3Si and R_3Ge groups, whereas the $[PO_2Cl_2]^{\ominus}$ ion (II) is favoured in compounds with big cations of little polarizing activity, such as the nitron und brucinium ions. It is interesting that in contrast to the difluorophosphate ion the dichlorophosphate group is able to form chelates, as shown by the examples of type (III) in Table 2. It is true that the comparatively wide OPO bond angle of 122° (found in $[Mn(PO_2Cl_2)_2 \cdot 2\,CH_3COOC_2H_5]_\infty$ (*77*) and in $[Mg(PO_2Cl_2)_2 \cdot 2\,POCl_3]_\infty$ (*73*)) which seems to hinder chelation, is practially the same as in $K[PO_2F_2]$ (*35*), but apparently it adapts very readily to smaller angles which are necessary for chelation. In $[Cl_3Sn(PO_2Cl_2) \cdot POCl_3]_2$ (type (IVa)) it decreases to 118° for example (*71*). The reason may be found in the ability of the P–Cl bond, to involve $(d_\pi\, d_\pi)$-orbitals, and thus to relieve the $(p_\pi\, d_\pi)$ bond, which induces a large OPO bond angle. Because of the energetically highly excited *d*-levels of the fluorine ligands, the P–F bond can do this only to a very small extent.

The most numerous structural types (IVa) and (IVb) of the dichlorophosphate group are related in so far as there is the same kind of linkage by the bifunctional

Table 2. Inorganic derivates of dichlorophosphoric acid, arranged in the order of their structural types

		Ref.
Typ I	$Cl_2(O)P{-}O{-}P(O)Cl_2$	*(61)*
	$[R_3P{-}O{-}P(O)Cl_2]^{\oplus}Cl^{\ominus}$	*(62)*
	($R = C_6H_5, C_4H_9, N(CH_3)_2$)	
	$(CH_3)_3Si{-}O{-}P(O)Cl_2$	*(63)*
	$(Oxinat)_2Si{-}[OP(O)Cl_2]_2$	*(64)*
	$Cl_4Sb(Pyridin){-}OP(O)Cl_2$	*(65)*
Typ II	$[Nitron]^{\oplus}[PO_2Cl_2]^{\ominus}$	*(66)*
	$[Brucinium]^{\oplus}[PO_2Cl_2]^{\ominus}$	*(66)*
	$[OxinH]^{\oplus}[PO_2Cl_2]^{\ominus}$	*(64)*
	$[Cl_4Sb(Pyridin)_2]^{\oplus}[PO_2Cl_2]^{\ominus}$	*(65)*
Typ III	$Tl(O_2PCl_2)_3$	*(67)*
	$Bi(O_2PCl_2)_3 \cdot POCl_3$	*(68)*
	$TiO(O_2PCl_2)_2$	*(40)*
	$VOCl_2(O_2PCl_2) \cdot POCl_3$	*(69)*
Typ IVa	$[(CH_3)_2Al(O_2PCl_2)]_3$	*(70)*
	$[(CH_3)_2Ga(O_2PCl_2)]_2$	*(70)*
	$[(CH_3)_2In(O_2PCl_2)]_2$	*(70)*
	$[Cl_3Sn(O_2PCl_2) \cdot POCl_3]_2$	*(71)*
	$[Cl_4Sb(O_2PCl_2)]_2$	*(65)*
	$[Cl_3Ti(O_2PCl_2) \cdot POCl_3]_2$	*(72)*
	$[Cl_3Zr(O_2PCl_2) \cdot POCl_3]_2$	*(72)*
Typ IVb	$Mg(O_2PCl_2)_2 \cdot 2\,POCl_3$	*(73)*
	$Ca(O_2PCl_2)_2 \cdot 2\,CH_3COOC_2H_5$	*(74)*
	$M(O_2PCl_2)_3$	*(75)*
	(M = Al, Ga, In, Fe)	
	$Si(O_2PCl_2)_4$	*(76)*
	$Mn(O_2PCl_2)_2 \cdot 2\,CH_3COOC_2H_5$	*(77)*
	$Mn(O_2PCl_2)_2 \cdot 2\,POCl_3$	*(78)*
	$Nd(O_2PCl_2)_3$	*(79)*
Typ V	$[MoO_2(O_2PCl_2) \cdot POCl_3]_{\infty}$	*(80)*

dichlorophosphate group. Whether polymerisation is achieved – at present the only known trimer is $[(CH_3)_2Al(O_2PCl_2)]_3$ *(70)* – is essentially dependent on coordinational points of view and the charge distribution of the structural partner. So the tetrachloroantimony group with the bifunctional dichlorophosphate group obtains a coordination number of 6 for the antimony atom by dimerization (A) *(65)*, whereas the trichlorotin group comes the same coordination number by additional solvatation with a $POCl_3$ molecule (B) *(71)*:

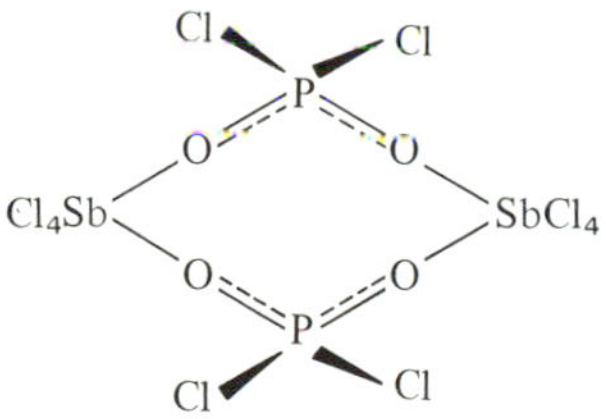

(A)

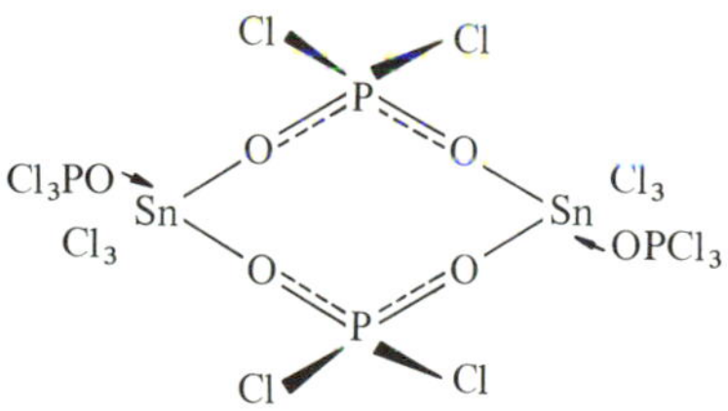

(B)

On the other hand the dichlorophosphates of bivalent metals, such as Mg, Mn^{II}, Co^{II} etc., are polymeric compounds of structural type (IV). The coordination number 6, which is typical for these metals, when oxygen is the ligand, is achieved by solvatation with $POCl_3$ or by the oxygen atoms of the keto group of $CH_3COOC_2H_5$ *(73, 77, 78)*. The solvent molecules, which are arranged in cis-position, are interchangeable:

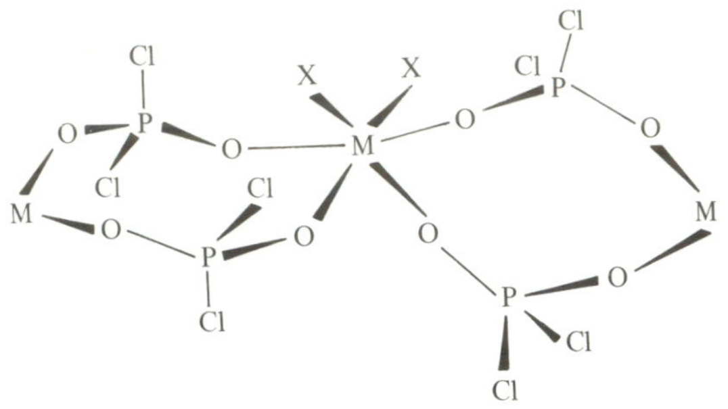

$M(O_2PCl_2)_2 \cdot 2\,X$

$M = Mg, Mn$
$X = POCl_3, CH_3COOC_2H_5$

The only example of the trifunctional type (V), known at the time, is the $MoO_2(PO_2Cl_2) \cdot POCl_3$, in which the molybdenum atom completes its octahedral sphere, in spite of abundant solvent, by aid of a chlorine atom of the bridging dichlorophosphate group (*80*):

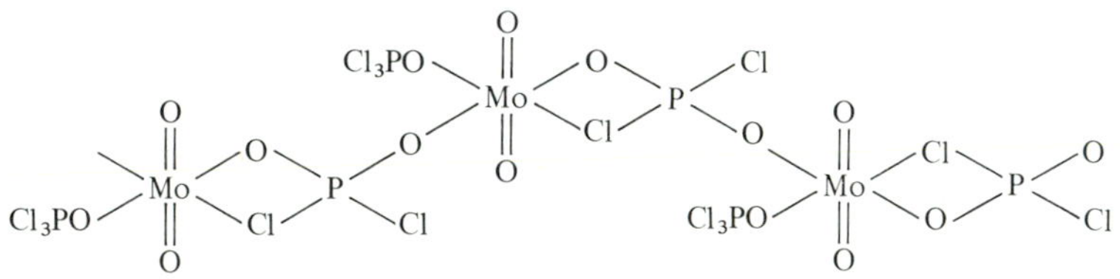

The oldest of the numerous *syntheses of dichlorophosphates* dates back to 1911 (*81*); it makes use of the reaction of metal oxides (MgO, CaO, MnO) with $POCl_3$ in presence of solvatating organic molecules (e.g. ethyl acetate = E). This reaction can be understood by the access of moisture from the air (26) as well as by the simultaneous formation of metal chloride (27):

$$CaO + 2\,POCl_3 + 2\,E + H_2O \rightarrow \text{“}CaO \cdot P_2O_3Cl_4 \cdot 2\,E\text{”} + 2\,HCl \qquad (26)$$

$$2\,CaO + 2\,POCl_3 + 2\,E \rightarrow \text{“}CaO \cdot P_2O_3Cl_4 \cdot 2\,E\text{”} + CaCl_2 \qquad (27)$$

However, the nature of these compounds could not be determined until in 1963 *Danielsen et al.* made a crystal structure analysis of the manganese compound (*77*). Later on these dichlorophosphates, in analogy to reaction (26), were prepared from dichlorophosphoric anhydride too:

$$MnO + P_2O_3Cl_4 + H_2O + 2\,E \rightarrow Mn(PO_2Cl_2)_2 \cdot 2\,E + 2\,HCl \qquad (28)$$

With the carbonates instead of the oxides as starting compounds (*74*) it is advisable to take the free dichlorophosphoric acid as reactant:

$$CaCO_3 + 2\ HPO_2Cl_2 + 2\ E \rightarrow Ca(PO_2Cl_2)_2 \cdot 2\ E + CO_2 + H_2O \qquad (29)$$

In a comparable manner some organometallic oxides, such as hexamethyl disiloxane (*63*) and -germane, react (*82*):

$$(CH_3)_3Si{-}O{-}Si(CH_3)_3 + POCl_3 \xrightarrow{110^\circ} (CH_3)_3Si[PO_2Cl_2] + (CH_3)_3SiCl \qquad (30)$$

Using the mixed ether with Si–O–Ge bridges as reagent, the preferred formation of $(CH_3)_3Si(PO_2Cl_2)$ is found (*82*).

Remarkable complex ions are formed in reactions of triphenylphosphine oxide with $POCl_3$ (*62*):

$$(C_6H_5)_3PO + POCl_3 \rightarrow [(C_6H_5)_3P{-}O{-}POCl_2]^{\oplus}Cl^{\ominus} \qquad (31)$$

Trialkylphosphine oxides and tris(dimethylamino)phosphine oxide are undergoing this reaction as well (*62*).

Even oxyhalides such as $SnOCl_2$, $TiOCl_2$ etc. react readily in boiling phosphoryl chloride, forming dichlorophosphates (*72, 83, 84, 85*):

$$2\ TiOCl_2 + 4\ POCl_3 \rightarrow [TiCl_3(PO_2Cl_2) \cdot POCl_3]_2 \qquad (32)$$

A great many of metal chlorides, solvatable by $POCl_3$, such as $SbCl_5$, $SnCl_4$, $TiCl_4$, $AlCl_3$, $FeCl_3$ etc., react with Cl_2O, which is easily prepared by passing Cl_2 over mercuric oxide (*65, 75*)

$$SbCl_5 \cdot POCl_3 + Cl_2O \xrightarrow{POCl_3} Cl_4Sb(PO_2Cl_2) + 2\ Cl_2 \qquad (33)$$

$$FeCl_3 \cdot 3\ POCl_3 + 3\ Cl_2O \xrightarrow{POCl_3} Fe(PO_2Cl_2)_3 + 6\ Cl_2 \qquad (34)$$

The driving force of these reactions is the redox compensation of the involved Cl-species, which are inversely polarized, possessing $Cl^{\oplus}$-character in Cl_2O and $Cl^{\ominus}$-character in the metal chloride and $POCl_3$ resp. (*67*).

The volatility of trifluoroacetyl chloride may explain the smooth reaction when preparing neodymium(tris)dichlorophosphate, which is of some interest for dye-lasers, from trifluoroacetate and $POCl_3$ (*79*):

$$Nd(CF_3COO)_3 + 3\ POCl_3 \rightarrow Nd(PO_2Cl_2)_3 + 3\ CF_3COCl \qquad (35)$$

In some cases the anhydrous acetates $M^{II}(CH_3COO)_2$ (M^{II} = Ni, Mn) can be used as starting compounds, according to Eq. (35), too; it is necessary however to employ excessive $POCl_3$ for the preparation of $M^{II}(PO_2Cl_2)_2 \cdot 2\ POCl_3$, in order to shift the equilibrium adequately (*78*).

Of some importance for the preparation of various dichlorophosphates are cleaving reactions of the P–O–P bridge in $P_2O_3Cl_4$, which quite often reacts completely with some organometallic oxides (36) (*82*) or suitably substituted ethers (37) (*63*):

$$(CH_3)_3Ge\text{–}O\text{–}Ge(CH_3)_3 + P_2O_3Cl_4 \xrightarrow{-20^\circ} 2\,(CH_3)_3Ge[PO_2Cl_2] \quad (36)$$

$$(CH_3)_3Si\text{–}O\text{–}R + P_2O_3Cl_4 \longrightarrow (CH_3)_3Si[PO_2Cl_2] + ROPOCl_2 \quad (37)$$

In a like way $P_2O_3Cl_4$ reacts with metal chlorides (*65, 86*), where at least with the *Lewis* acids among them, a primary formation of adducts, e.g. $SbCl_5 \cdot P_2O_3Cl_4$, is likely (*87, 65*):

$$4\,SbCl_5 + 2\,P_2O_3Cl_4 \rightarrow [SbCl_4(PO_2Cl_2)]_2 + 2\,SbCl_5 \cdot POCl_3 \quad (38)$$

$$Sn(CH_3)_2Cl_2 + 2\,P_2O_3Cl_4 \rightarrow Sn(CH_3)_2[PO_2Cl_2]_2 + 2\,POCl_3 \quad (39)$$

Also, the reaction of metal chlorides with anhydrous dichlorophosphoric acid is convenient for the preparation of some dichlorophosphates, as is shown by the following examples (*88, 64, 74, 76*):

$$(CH_3)_3SiCl + HPO_2Cl_2 \longrightarrow (CH_3)_3Si[PO_2Cl_2] + HCl \quad (40)$$

$$Cl_2Si(oxinat)_2 + 2\,HPO_2Cl_2 \longrightarrow [PO_2Cl_2]_2Si(oxinat)_2 + 2\,HCl \quad (41)$$

$$CaCl_2 + 2\,HPO_2Cl_2 + 2\,ester \longrightarrow Ca(PO_2Cl_2)_2 \cdot 2\,ester + 2\,HCl \quad (42)$$

$$SiCl_4 + 5\,HPO_2Cl_2 \xrightarrow{ether} [(C_2H_5)_2OH]^{\oplus}[Si(PO_2Cl_2)_5]^{\ominus} \quad (43)$$

Reactions with the trialkyls of aluminium, gallium and indium are proceeding without complications as well (*39*):

$$MR_3 + HPO_2Cl_2 \rightarrow R_2M[PO_2Cl_2] + RH \quad (44)$$

Last, but not least it should be mentioned, that the acidity of dichlorophosphoric acid is sufficient, to dissolve some metals (Fe, Co, Ni) under formation of hydrogen and dichlorophosphates (*78*):

$$Fe + 2\,HPO_2Cl_2 \rightarrow Fe(PO_2Cl_2)_2 + H_2 \quad (45)$$

3.6. Dibromophosphates $[PO_2Br_2]^{\ominus}$

In comparison to the chloro compounds the dibromophosphates are by far less thoroughly investigated. *Grunze et al.* were the first to succeed in the isolation of the nitron salt of dibromophosphoric acid, by partial hydrolysis of $POBr_3$ in cooled aceton solution (*8*). Later on it could be shown, that even CaO and MgO form solvatated dibromophosphates, when added to a solution of $POBr_3$ in ethyl acetate and hydrolized partially (*89, 74*):

$$MgO + 2\,POBr_3 + H_2O + 2\,E \rightarrow Mg(PO_2Br_2)_2 \cdot 2\,E + 2\,HBr \qquad (46)$$

In analogy to reaction (35), neodymium(tris)dibromophosphate, too, could be prepared from trifluoroacetate and $POBr_3$ (*79*).

The dibromophosphates of Mg, Ca and Nd are polymeric compounds, corresponding to the structural type (IVb) of the dichlorophosphates. In contrast the way to $[Br_3Ti(O_2PBr_2)POBr_3]_2$ (*72, 85*), which is analogous to Eq. (32), consists in a linkage of the PO_2Br_2 group corresponding to (IVa).

3.7. *Mixed dihalophosphates* $[PO_2FCl]^{\ominus}$, $[PO_2FBr]^{\ominus}$

These two dihalophosphates are only known in form of their lithium salts, which are obtained from the etheric solutions of the free acids (see section 2.) by addition of lithium bromide (*7*). Their structures are not known in detail, yet their infrared spectra do not show any remarkable characteristics, so most certainly these compounds are of ionic character.

4. Structural investigations

The main investigation methods, to give information about structure and bonding in the halophosphates, are vibrational and NMR spectroscopy, which give a fairly complete picture. Up to now, X-ray studies are restricted to a few examples only; indirectly they are of some importance in the discussion, especially of infrared and Raman spectra however. In some cases indirect information has been instructive, e.g. Mössbauer spectra of iron and tin halophosphates, so that these measurement shall be mentioned too.

4.1. Crystal structure analyses

Up to know crystal structure analyses have been made with difluorophosphates, monofluorophosphates and dichlorophosphates only. Table 3 gives the bond lengths and bond angles in these three halophosphates.

Table 3. Crystal structures of halophosphates

Compound	r_{PO} [Å]	r_{PX} [Å]	∢ OPO [°]	∢ XPX [°]	∢ XPO [°]	Ref.
$K[PO_2F_2]$ [a])	1,470 (5)	1,575 (5)	122,4 (4)	97,1 (4)	108,6 (3)	*(35)*
$NH_4[PO_2F_2]$	1,457 (8)	1,541 (7)	118,7 (5)	98,6 (5)	109,2 (5) 109,6 (5)	*(94)*
$Ca[PO_2F_2]_2 \cdot 2\,E$ [b])	1,41	1,50	126	95	108	*(47)*
$Ca[PO_3F] \cdot 2\,H_2O$	1,506 (1) 1,515 (1) 1,503 (2)	1,583 (1)	115,64 (8) 116,48 (7) 109,52 (8)		102,68 (6) 105,03 (6) 106,13 (7)	*(26)*
$(NH_4)_2[PO_3F] \cdot H_2O$	1,508 (1) 1,506 (1) 1,501 (1)	1,586 (1)	112,95 (7) 114,53 (7) 114,57 (8)		104,31 (7) 104,05 (6) 104,88 (7)	*(26)*
$Sn[PO_3F]$	1,46 (4) 1,48 (5) 1,51 (8)	1,57 (3)	–	–	–	*(33)*
$Mn[PO_2Cl_2]_2 \cdot 2\,E$ [b])	1,46 1,47	2,01 2,02	122,0	101,7	107,7 106,2 108,2 109,3	*(77)*
$Mg[PO_2Cl_2]_2 \cdot 2\,POCl_3$	1,440 (10) 1,434 (12)	2,005 (8) 1,995 (7)	122,0 (6)	100,8 (3)	108,0 (7) 108,6 (7)	*(73)*
$[SnCl_3[PO_2Cl_2] \cdot POCl_3]_2$	1,485 (12) 1,498 (12)	1,988 (7) 1,954 (1)	118,0 (7)	103,6 (3)	106,9 (7)	*(71)*

[a]) $RbPO_2F_2$ and $CsPO_2F_2$ have comparable structural parameters;
[b]) E = ethyl acetate

In the three isostructural alkaly difluorophosphates KPO_2F_2 (*35*), $RbPO_2F_2$ and $CsPO_2F_2$ (*90*), within the error limit, only one PO and PF distance is found (see Table 3), so that the $[PO_2F_2]^{\ominus}$ symmetry is C_{2v}. On the other hand the PO and PF bond lengths are clearly different, as well as the bond angles OPO (122.4°) and FPF (97.1°), indicating considerable PO-π-bonding (see section 3.). The sum of the P and O covalent radii according to Pauling is 1.76 Å (*91*), according to the *Schomaker-Stevenson* bond length it amounts to 1.71 Å (*92*); with 1.47 Å the experiment gives a significantly different value. With the relation of *Robinson* (n_{PO} = bond order of the PO bond)

$$n_{PO} = 23.8/r_{PO}^{8} + 0,74$$

$n_{PO} = 1.83$ is calculated for the $[PO_2F_2]^{\ominus}$ ion, which means 83% π-bonding (*35*). Similar results are obtained from the vibrational spectra (see section 4.2.). These calculations allow 17% π-bonding for the PF bond, which again is confirmed by the comparison of the P and F covalent radii (1.74 Å acc. to *Pauling*, 1.65 Å according to (*92*)) with the experimental value of 1.575 Å (*35*).

The structure of $NH_4[PO_2F_2]$ (*94*) resembles the structures of the alkali difluorophosphates, but remarkable differences arise as the consequence of N–H ··· O and N–H(O)(O)-hydrogen bridges. They become evident in the change of the bond angles of the $[PO_2F_2]^{\ominus}$ ion, which decrease to 118.7° for the OPO group and increase to 98.6° for the FPF-group. Obviously the π-bonding in the PO bonds has been weakened by the electrostatic interaction with the NH protons, which in turn has increased the PF-π-bonding, because the F-atoms are not involved in hydrogen bridging (*94*).

The structural analysis of $Ca[PO_2F_2]_2 \cdot 2$ ethyl acetate (*47*) has not been done in a very detailed way, yet it is to note, that the solvate molecules, with their keto oxygen atoms, are occupying the trans positions of the calcium ion, which is surrounded octahedrally by 6 O-atoms. In contrast the solvate molecules take the cis-position in the similarly solvatated dichlorophosphates $Mn(PO_2Cl_2)_2 \cdot 2$ ethyl acetate (*77*) and $Mg(PO_2Cl_2)_2 \cdot 2\ POCl_3$ (*73*), which belong to the polymeric structural type (IVb) (s. page 62) as well as the calcium compound. It is likely, that for this behaviour the packing of these dihalophosphates, which are of different size, plays an important role.

Looking at the crystal structures of the monofluorophosphates, it is surprising, that the three PO bonds are of different lengths. These differences are out of the limits of error at least in case of the very exactly calculated data of $Ca[PO_3F] \cdot 2\ H_2O$ and $(NH_4)_2[PO_3F] \cdot H_2O$ (*26*) (see Table 3). In fact this leads to a C_1 symmetry for the $PO_3F^{\ominus}$ ion.

In both compounds the hydratemolecules form hydrogen bridges, in which as in the difluorophosphates, only the O-atoms of the $[PO_3F]^{2\ominus}$ ions participate; in $(NH_4)_2[PO_3F] \cdot H_2O$ there are additional N–H ··· O bridges (*26*). In the $PO_3F^{2\ominus}$ ion the formal bond order n_{PO} decreases from 1.83 to 1.67 (*95*) in comparison to

$[PO_2F_2]^{\ominus}$, which indicates single bond character for the P–F bond. Actually the difference between the calculated P–F bond length (see above) and the experimental data (1.583 Å in $Ca[PO_3F] \cdot 2\,H_2O$ (*26*)) diminishes, yet remains significant. Although the P–F stretching vibration is shifted to longer wavelengths with respect to $[PO_2F_2]^{\ominus}$ (*96*), the conclusion of negligible π-bonding should be drawn only with adequate reserve. The relations between $[PO_3F]^{2\ominus}$ and the isoelectronic $[SO_4]^{2\ominus}$ ion have already been pointed out in section 3.1.

Besides the crystal structure data shown in Table 3, the lattice constants of various alkali monofluorophosphate have been determined (*97, 98, 99, 100*). A very similar structure of $Pb[PO_3F]$ and $Pb[SO_4]$ has been concluded from their isostructurality (*101*). The structural investigations of dichlorophosphates are confined to the "linking" structural types (IVa) and (IVb) of the dichlorophosphate group, so that at present comparisons with other structural types are only possible with the aid of vibrational spectroscopy. The OPO bond angle of 122°, found in the dichlorophosphates of magnesium (*73*) and manganese (*77*), is practically the same as in the difluorophosphates. It is not possible to state comparable π-bonding from this fact alone, because the different volume of the chlorine ligands has to be regarded too. Actually the relation of *Robinson* (see above) yields $n_{PO} = 1.64$ for the PO bond order of the $PO_2Cl_2^{\ominus}$ group in $Mg(PO_2Cl_2)_2 \cdot 2\,POCl_3$, which means a significant reduction of π-bonding in comparison to $K[PO_2F_2]$ ($n_{PO} = 1.83$), and a correspondingly increased double bond character for the P–Cl bonds. In $[Cl_3Sn(PO_2Cl_2) \cdot POCl_3]_2$ n_{PO} of the $PO_2Cl_2^{\ominus}$ group again decreases to 1.38, which is in accordance with the smaller bond angle of 118°. These differences in bonding become evident in the vibrational spectra (see below) too: An appreciable amount of π-bond character is equivalent to high PO_2 stretching frequencies, while a shift of these bands to lower frequencies indicates decreasing PO bond strengths. As the PO_2 stretching vibrations in the tin compound $[Cl_3Sn(PO_2Cl_2) \cdot POCl_3]_2$ (*71*) and the antimony compound $[Cl_4SbPO_2Cl_2)]_2$ (*65*), which are very similar in structure, are of almost the same frequency, it is likely, that the bond angles and bond lengths are of comparable order of magnitude.

4.2. Vibrational Spectra

In the undisturbed state, the halophosphates can be classified by the following point groups (vibrational species in brackets):

$[PO_3F]^{2\ominus}$	$[PO_2X_2]^{\ominus}$	$[POSX_2]^{\ominus}$	$[PO_2XY]^{\ominus}$	$[POSXY]^1$
C_{3v}	C_{2v}	C_s	C_s	C_1
(E, A_1)	(A_1, B_1, B_2)	(A', A'')	(A', A'')	(A)

For each of the pentatomic ions 9 fundamental vibrations are expected, which are active in both spectroscopic methods (i.r. and *Raman*) only in case of the point groups C_s and C_1. For C_{3v} the number of bands is decreased to 6 on account of double degeneration, in C_{2v} one deformation of the species A_2 is only Raman-active. All complex ions are expected to show 4 stretching vibrations, except for the point group C_{3v}, where this number is reduced to three, because one stretching mode ist doubly degenerate (species E). In any case the selection rules require all stretching vibrations to be active in both effects (i.r. and *Raman*).

The discussion shall be confined to the particularly characteristic and in general authentically known stretching vibrations, as these are exceedingly sensitive to changes in bonding. In the non ionic derivatives of the halophosphates, of course, the symmetry of the molecule may be considerably different from that of the ion; so for the dimeric species the number of observable vibrations of the halophosphate group is doubled on account of the existence of in phase and out of phase vibrations. However in many cases the assignment can be done assuming local symmetries, based on the point groups given above.

In Table 4 the stretching frequencies of a number of halophosphates are listed. Of special interest is the discrimination of the different structural types (I–V) discussed for dichlorophosphates in section 3.5., by means of vibrational spectroscopy. Generally it is to note, that a rough conclusion can be drawn from the PO_2 stretching frequencies, as the O-atoms, because of their direct bonding, are in particular subject to structural changes. The most striking difference is found between types (I) and (II). In (I) the fixation of one P–O–σ bond brings along a PO double bond with terminal character, so that the two PO stretching frequencies are externely different (wavenumbers in cm^{-1}):

PyCl₄Sb–O (1000) P (1240) =O; Cl, Cl

(I) [65]

$[O \cdots P(Cl)_2 \cdots O]^{\ominus}$ (1135, 1100) $[SbCl_4 \cdot 2\ Py]^{\oplus}$

(II) [65]

On the other hand, in the ion (II) the PO_2 stretching vibrations move towards each other quite closely because of "complete resonance, as is obvious from the examples $[SbCl_4 \cdot 2\ Py]^{\oplus}[PO_2Cl_2]^{\ominus}$ (*65*), $Mg(PO_2Cl_2)_2 \cdot 2\ POCl_3$ (1145, 1092) (*78*), $Mn(PO_2Cl_2)_2 \cdot 2\ POCl_3$ (1129, 1092) (*78*) and $Fe(PO_2Cl_2)_2 \cdot 2\ POCl_3$ (1135, 1095) (*78*). Intermediate positions are taken by the structural types (III–V), as indicated by the PO_2 stretching vibrations:

(III)	$Cl_2VO(PO_2Cl_2) \cdot POCl_3$	1233, 1085 cm^{-1}	$\Delta\nu$ = 148 cm^{-1}	(*69*)
(IVa)	$[Cl_4Sb(PO_2Cl_2)]_2$ [2])	1136, 1065 cm^{-1}	$\Delta\nu$ = 71 cm^{-1}	(*65*)
(IVb)	$[Nd(PO_2Cl_2)_3]_\infty$ [2])	1238, 1090 cm^{-1}	$\Delta\nu$ = 148 cm^{-1}	(*102*)
(V)	$MoO_2(PO_2Cl_2) \cdot POCl_3$	1178, 1080 cm^{-1}	$\Delta\nu$ = 98 cm^{-1}	(*80*)

2) average values of frequencies, see Table 4

Table 4. Vibrational spectra of halophosphates (frequencies in cm^{-1})

Compound	Spectr. methodes	$\nu_{as}PO_n$	ν_sPO_n	νPS	$\nu_{as}PX_n$	ν_sPX_n	Notes	Ref.
$K_2[PO_3F]$	IR, RE	1170 (E)	1008 (A_1)			705 (A_1)	Raman spectra in the melt 700 °C	(96)
$Ba[PO_3F]$	IR	1163, 1127	1005			745 (broad)		(105)
$K[PO_2F_2]$	IR, RE	1311 (B_1)	1145 (A_1)		857 (B_2)	834 (A_1)	Raman spectra in the melt 300 °C	(96)
$[(CH_3)_2Al(PO_2F_2)]_3$	IR, RE	1312, 1290	1210		980	931	molecular symmetry D_3	(70)
$[(CH_3)_2Ga(PO_2F_2)]_2$	IR	1295, 1262	1171		949	899	molecular symmetry D_2	(70)
	RE	1300, 1263	1187, 1175		952	905	(In-compound known, too)	
$In(PO_2F_2)_3$	IR	1269	1179		962	910		(40)
$(CH_3)_3Si(PO_2F_2)$	IR	1358 ($\nu P=O$)	1110 ($\nu_{as}SiOP$)		940	910		(38)
$[(CH_3)_2Sn(PO_2F_2)_2]_\infty$	IR, RE	1258 (B_1)	1168 (A_1)		928 (B_2)	888 (A_1)	methyl groups in transposition; assignment acc. to local symmetry C_{2v}. Compounds with higher alkyl substance are known, too	(43)
$Xe(PO_2F_2)_2$	RE	1299	1123		970	890		(44)
$[Cl_2Ti(PO_2F_2)_2]_\infty$	IR, RE	1215	1155		972	922	νTiO 435, 447 νTiCl 415, 431	(50)
$P_2O_3F_4$	RE	1390, 1370 ($\nu P=O$) 987 ($\nu_{as}POP$)	480 (ν_sPOP)		951, 890 855			(61)
$(HPO_2F_2)_n$	RE	1335 ($\nu P=O$)	1070 ($\nu P-O$)		978	888	Compound is associated over P=O–H–O–P bridges	(13)
$Fe(PO_2F_2)_3$	IR	1242	1173		965	914	493 νFeO_6 (F_{1u})	(40)
	RE	1236	1170		965	919	450 νFeO_6 (A_{1g})	(41)

Table 4. (continued)

$[N(C_2H_5)_4]^{\oplus}[POSF_2]^{\ominus}$	IR		1218	658				(57)
$(CH_3)_3Si(OPSF_2)$	IR		1085 (ν_{as}SiOP)	760	930	910		(58)
$[P(C_6H_5)_4]^{\oplus}[POSFCl]^{\ominus}$	IR		1230	638		795 (νPF) 483 (νPCl)	Tetraphenylarsonium salts are described also	(22)
$[P(C_6H_5)_4]^{\oplus}[POSCl_2]^{\ominus}$	IR		1215	645	462	437		(22)
$[Be(PO_2Cl_2)_2]_\infty$	IR	1259	1147		708	590		(75)
$[Mg(PO_2Cl_2)_2 \cdot 2\,POCl_3]_\infty$	IR	1145	1092		554	536	νPO($POCl_3$): 1290, breit	(78)
$[Al(PO_2Cl_2)_3]_\infty$	IR	1258	1147		630	587	The structurally analogous compounds of Ga, In and Fe^{III} are also known (75)	(75)
$Tl(PO_2Cl_2)_3$	IR	1185	1080		600	555	420 νTlO_6 (F_{1u}) 160 δTlO_6 (F_{1u}) (PO_2Cl_2 groups are linked as chelates)	(67)
$[(CH_3)_2Al(PO_2Cl_2)]_3$	IR, RE	1275, 1230	1156 (only RE), 1138		618	613	Molecular symmetry D_3 The diethyl compound is dimeric (70)	(70)
$[(CH_3)_2Ga(PO_2Cl_2)]_2$	IR, RE	1270, 1230	1124 (only RE), 1110		597	585 (RE)	Dimethyl derivatives of In and Tl are known	(70)
$[Cl_3Sn(PO_2Cl_2)POCl_3]_2$	IR RE	1190 1163	1089, 1066 1084		636, 615 629	589, 565, 532, 582, 527	Molecular symmetry C_{2h}. νP=O ($POCl_3$): 1269, 1240 (IR) 1202 (RE)	(72)
$[Cl_4Sb(PO_2Cl_2)]_2$	IR RE	1165 1108	1058 1072		636 628, 600	584	Structure with a centre of symmetry (D_{2h}, C_{2h})	(65)
$[Cl_4Sb \cdot Pyridin(PO_2Cl_2)]$	IR	1240 (νP=O)	1000 (νP – O)		570	570	Structural type (I)	(65)
$[Cl_4Sb(Pyridin)_2]^{\oplus}[PO_2Cl_2]^{\ominus}$	IR	1135	1100		600	550		(65)

Table 4. (continued)

$Bi(PO_2Cl_2)_3 \cdot POCl_3$	IR, RE	1208, 1180 (IR)	1070 (IR) 1055 (IR)	585	555	PO_2Cl_2 groups linked as chelates; hexahedral environment of Bi	(*68*)
$[Cl_3Ti(PO_2Cl_2) \cdot POCl_3]_2$	IR	1200	1089	613	580	$\nu PO(POCl_3)$: 1243	(*72*)
$[TiO(PO_2Cl_2)_2]_\infty$	IR	1200	1075	630	592	Linked by Ti=O–Ti chains, νTiO: 972	(*40*)
$[Cl_3Zr(PO_2Cl_2) \cdot POCl_3]_2$	IR	1185	1083	632	586	$\nu PO(POCl_3)$: 1230	(*72*)
$Cl_2VO(PO_2Cl_2) \cdot POCl_3$	IR	1233	1085	605	590	$\nu PO(POCl_3)$: 1270 νVO: 1009	(*69*)
$[Cl_2V(PO_2Cl_2)_2]_\infty$	IR	1206	1094	612	579	νVO: 505	(*69*)
$Mn(PO_2Cl_2)_2 \cdot 2\,E$ [a])	IR	1124	1093	573	552	$\nu C{=}O$ (E) [a]): 1700 (A_1), 1688 (B_1)	(*78*)
$[MoO_2(PO_2Cl_2) \cdot POCl_3]_\infty$	IR	1178	1080		490	$\nu PCl \cdots Mo$: 368, νMoO: 995, $\nu PO(POCl_3)$: 1268	(*80*)
$Nd(PO_2Cl_2)_3$	IR	1245, 1230	1090	615, 585	550	νNdO: 410, 390 approximate symmetry D_3 at the Nd	(*102*)
$P_2O_3Cl_4$	RE	1312 ($\nu P{=}O$), 962 ($\nu_{as}POP$)	425 ($\nu_s POP$)	611, 560	531, 507		(*61*)
$(HPO_2Cl_2)_2$	RE	1259 ($\nu P{=}O$)	986 (P – O)	595	554	νOH: 2819	(*20*)
$[Br_3Ti(PO_2Br_2) \cdot POBr_3]_2$	IR	1153	1074	500	485	$\nu PO(POBr_3)$: 1217	(*72*)
$Nd(PO_2Br_2)_3$	IR	1230	1090	520	490	νNdO: 415, 370 (approximate symmetry D_3 at the Nd)	(*102*)
$Li(PO_2FCl)$	IR	1254	1135	876 (νPF)	630 (νPCl)		(*7*)
$Li(PO_2FBr)$	IR	1282	1160	874 (νPF)	558, 522		(*7*)

[a]) E = ethyl acetate

Hence the realisation of one of these structural types can not be verified from the PO stretching vibrations alone; additional information is necessary. In general, the knowledge of the molecular weight is important for the decision, as the chelates (III) form monomeric, the bridging types (IVa) dimeric, and the types (IVb) and (V) polymeric species. In the eight-ring compounds ring coupling causes the appearance of four PO_2 stretching frequencies, which are to be interpreted as in- and out of phase vibrations of symmetrical and antisymmetrical species. In rings of high symmetry (C_{2h}, D_{2h}) these are subject to the mutual exclusion rule, which permits direct statements about the ring geometry. The polymeric compounds of type (IVb), which contain 8-rings too, exhibit metal-oxygen-stretching modes – in contrast to the also polymeric ionic types (II), which do not display such vibrations.

This could be shown for $Nd(PO_2Cl_2)_3$ and $Nd(PO_2Br_2)_3$ by observation of Nd–O stretching bands in the region of about 400 cm^{-1}; thus these compounds were doubtlessly attributed to structural type (IVb) (*102*). The use of this criterion requires the knowledge of the vibrational spectra in the far i.r. region however. In this context it is of some interest, that $Fe(PO_2Cl_2)_2 \cdot 2\,POCl_3$ with divalent iron belongs to the ionic type (II) (*78*), whereas the appearance of νFeO_6 at 460 cm^{-1} (A_{1g}, *Raman*) and 492 cm^{-1}(F_{1u}, i.r.) resp. in the dichlorophosphate of Fe^{III} indicates the polymeric structural type (IVb) (*41*). As the PCl_2 stretching vibrations, except for type (V), are of purely terminal character, they are of little or no importance for the structural assignment. Moreover they are observed in the relatively narrow range between 550 to 630 cm^{-1}. However this fact allows to tell whether type (V) is existent or not, for in this structure, besides a terminal P–Cl bond, there is another bridged one which gives rise to a significant red shift. So in $MoO_2(PO_2Cl_2) \cdot POCl_3$ $\nu PCl_{term.}$ is found at 490 cm^{-1}, while νPCl_{bridge} appears at 368 cm^{-1} (*80*). The dichlorophosphates, which are solvatated by $POCl_3$-molecules in many cases, show a very characteristic PO stretching band at 1230 to 1260 cm^{-1} ($POBr_3$ at about 1220 cm^{-1}), which allows the definite identification of POX_3 solvates. As a consequence of the bonding activity by the O-atom this band is shifted to lower frequencies in comparison to free $POCl_3$ (1295 cm^{-1}) (*103*).

An analogous behaviour is found for the C=O stretching vibration in solvatated ethyl acetate. In $Mn(PO_2Cl_2)_2 \cdot 2\,E$ the crystallographically determined cis-position of the solvent molecules (*77*), which are coordinated with their keto O-atoms, can be realised by the double band at 1700/1688 cm^{-1}.

It can be understood as one A_1 and one B_1 mode, assigned on the basis of a local symmetry C_{2v} (*78*). In comparison to free ethyl acetate, the CO stretching frequency of which lies at 1742 cm^{-1} (*104*), this band is shifted by about 50 cm^{-1} by Mn–O bonding.

Analogous spectroscopic behaviour, as pointed out for the dichlorophosphates for example, can be expected in the case of the difluorophosphates and the dihalophosphates with two different halogen atoms.

In the monofluorophosphate ion the PO stretching vibrations (1170 (E) and 1008 cm^{-1} (A_1)) are of lower frequency compared with the difluorophosphate ion (1311 (B_1), 1145 cm^{-1} (A_1)), which is a result of the decrease of π-bonding in the

PO bond. The lower symmetry of the $[PO_3F]^{2\ominus}$ ion in the lattices of $Ca[PO_3F] \cdot 2\,H_2O$, $(NH_4)_2[PO_3F] \cdot H_2O$ (*26*) and $Sn[PO_3F]$ (*33*), which has been ascertained by crystallographic methods, allows three PO stretching modes for the point group C_1 of this ion. In the solid phase i.r. spectra of various fluorophosphates of alkali- and alkaline earth metals this in fact realised by an appropriate multiplicity of the PO stretching band; however occasionally the only effect is a striking broadening of the bands (*105*). On the other hand, in the Raman spectrum of a melt of K_2PO_3F C_{3v} symmetry of the anion is observed (*96*).

In case of the halothiophosphates, e.g. $[POSF_2]^{\ominus}$, the problem is, whether they are bonded to groups of high anisotropy via the O- or S-atom, or if actually both chalkogen atoms are to be considered for bonding. In trimethylsilyl thiodifluorophosphate there is spectroscopic evidence for structure (A) (*58*):

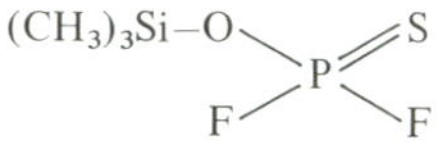

$(CH_3)_3Si{-}O{-}P({=}S)F_2$ (A) $\qquad$ $(CH_3)_3Si{-}S{-}P({=}O)F_2$ (B)

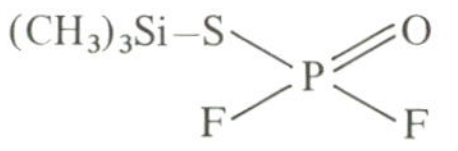

This becomes clear by the frequency of the PO stretching vibration at 1084 cm^{-1}, which is much too low for a terminal PO group as in (B) and thus is best characterised by "ν_{as}SiOP". On the other hand a band at 760 cm^{-1}, identified as PS stretching mode, can only be understood as a terminal one, because it is observed at even higher frequency than in the PSF_3 molecule (695 cm^{-1}) (*106*). In the ions $[POSF_2]^{\ominus}$ and $[POSCl_2]^{\ominus}$ νPS as well as νPO are shifted to longer wavelengths compared with the corresponding vibrations in POX_3 and PSX_3 (X = F, Cl) to about the same extent (see Table 4). This is certainly a consequence of the influence of the negative charge and the decreased π-bonding in the phosphorus halogen bond respectively.

4.3. NMR spectra

Table 5 shows the most important NMR data of halophosphates. In the species containing fluorine, apart from the chemical shifts of the ^{31}P-spectra, the ^{19}F-spectra and the coupling constants J_{PF} are available for the discussion of bonding. The importance of NMR spectroscopy for purity control, for equilibrium measurement, as kinetical method in the investigation of reaction processes and for the identification of unstable compounds shall not be discussed here, though these applications are of great importance for the halophosphates too.

On the basis of extensive spectroscopic experience (*107, 108*), it can be stated, that the substitution by ligands of higher electronegativity causes the high field chemical shift of the ^{31}P-spectra. However, this only holds if the structure and bonding of the molecules remain essentially unchanged, as can be anticipated for $P_2O_3Cl_4$ (δP = 10 ppm) (*109*)/$P_2O_3F_4$ (δP = 22.1 ppm) (*108*) or $[(CH_3)_2Ga(PO_2Cl_2)]_2$

Table 5 NMR Data of halophosphates

Compound	^{31}P δ [ppm]	^{19}F δ [ppm]	J_{P-F} [Hz]	J_{F-P} [Hz]	Solvent	Standard*)	Ref.
$P_2O_3Cl_4$	10,0				liquid	A	*(109)*
$[(C_2H_5)_2Ga(PO_2Cl_2)]_2$	– 1,2				CCl_4	A	*(39)*
$[(CH_3)_2Al(PO_2Cl_2)]_3$	1,4				CCl_4	A	*(39)*
$[(CH_3)_2Ga(PO_2Cl_2)]_2$	– 1,7				C_6H_6	A	*(39)*
$[(CH_3)_2In(PO_2Cl_2)]_2$	1,1				$(C_2H_5)_2O$	A	*(39)*
$[HPO_2Cl_2]_2$	– 9,5				liquid (~ 93%)	A	*(115)*
$[H_2PO_3F]_n$	7	– 11	937	939	H_2O	A, B	*(17)*
H_2PO_4F	2,1		1033		H_2O/H_2O_2	A, B	*(11)*
Na_2PO_3F	– 1,7	73,0	863		H_2O	A, C	*(108)*
Na_2PO_3F	0	– 7	841	805	H_2O	A, B	*(17)*
$NaHPO_3F$	4	– 6	910	911	H_2O	A, B	*(17)*
K_2PO_3F	– 2	– 9	883	964	H_2O	A, B	*(17)*
$KHPO_3F$	5	– 3	896	874	H_2O	A, B	*(17)*
$(NH_4)_2PO_3F$	– 4	– 8	891	922	H_2O	A, B	*(17)*
$NH_4PO_2F_2$	14,7	82,1	960		H_2O	A, C	*(108)*
HPO_2F_2	22,3	8,8	980		H_2O	A, C	*(11)*
$P_2O_3F_4$	22,1	85,7	980		liquid	A, C int.	*(108)*
HPO_3F_2	19,7	15,6	1064	1065	H_2O/H_2O_2	A, B	*(11)*
$(CH_3)_3Si-OPSF_2$		41,3		1115	$CFCl_3$	C int.	*(58)*
$(CH_3)_3Sn-OPSF_2$		37,1		1126	CCl_4	C	*(42)*
$[(C_6H_5)_4P]^{\oplus}[POSFCl]^{\ominus}$		43,6	1056		CH_3OH	A, C	*(22)*
HPO_2FCl	1,96	– 33,3	1050	1055	liquid	A, B	*(7)*
HPO_2FBr	15,9	– 46,3	1103	1106	liquid	A, B	*(7)*
$NaPO_2F_2$	14,7	82,0	959		H_2O	A, C	*(108)*
KPO_2F_2	14,8	82,2	960		H_2O	A, C	*(108)*
$[(CH_3)_2Al(PO_2F_2)]_3$	33,6	8,6	960	960	CCl_4	A, B	*(39)*
$[(CH_3)_2Ga(PO_2F_2)]_2$	29,6	6,1	960	960	$(C_2H_5)_2O$	A, B	*(39)*
$[(CH_3)_2In(PO_2F_2)]_2$	22,8	4,1	967	968	CCl_4	A, B	*(39)*
$[(CH_3)_2Tl(PO_2F_2)]_2$	16,5	3,0	960	959	Pyridine	A, B	*(39)*
$(CH_3)_3Sn(PO_2F_2)$		82,4		966	CCl_4	C	*(42)*
$(C_4H_9)_3Sn(PO_2F_2)$	25,5	82,3		966	liquid	A, C	*(38)*
$(CH_3)_3Si(PO_2F_2)$	28,7	82,8	1000	984	liquid	A, C	*(38)*

[a]) A = 85% H_3PO_4, external; B = CF_3COOH, external; C = $CFCl_3$, external.

($\delta P = -1.7$ ppm)/$[(CH_3)_2Ga(PO_2F_2)]_2$ ($\delta P = 29.6$ ppm) (*39*). On the other hand the change in the chemical shift, passing from PSF_3 ($\delta P = -32.0$ ppm) (*108*) to POF_3 ($\delta P = 35.5$ ppm) (*110*), can not be explained by the higher electronegativity of the O-ligand alone, but mainly by the higher fraction of PO π-bonding (*108*). A similar influence has to be assumed when comparing the ^{31}P–NMR spectra (see Table 5) of $[PO_3F]^{2\ominus}$ ($\delta P = -2$ ppm) and $[PO_2F_2]^{\ominus}$ ($\delta P = 15$ ppm). Bond lengths and -angles, as found by crystallographic methods (see section 4.1.) and the vibrational spectra (see section 4.2.) indicate a change of π-bonding in the PO bond. So in any case both, electronegativity as well as bonding, have to be considered. So the result of substituting the OH-ligand by the higher electronegative OOH-group, for example, is a low field shift of the ^{31}P triplet from 22.3 to 19.7 ppm (*11*), which may originate in a decrease of π-bonding in the PO bond. The variation of the alkali ions does not produce considerable changes in the chemical shifts of $[PO_3F]^{2\ominus}$ ions; solutions of the acid salts M^IHPO_3F show chemical shifts lying between those of the free ions and the acid (*17*).

In the series of the dimethylmetal difluorophosphates from aluminium to thallium, the chemical shift in the ^{31}P-spectra of $[(CH_3)_2Tl(PO_2F_2)]_2$ is most striking. In contrast to other compounds it is with 16.5 ppm quite close to the ionic difluorophosphates. Here the strongly solvatating solvent pyridine must be taken into account, which favours the formation of $(CH_3)_2Tl^{\oplus}$ and $[PO_2F_2]^{\ominus}$ respectively.

A remarkable splitting of the two signals in the ^{19}F-spectra of $P_2O_3F_4$ is observed, when traces of difluorophosphoric acid are carefully excluded and the spectrum is scanned at very slow speed (*111*). The doublet is split into two triplets symmetrical to each other with the intensity ratio 1 : 2 : 5, which is in accordance with the AA′XX′A″ A‴ expectation spectrum (*111*). The ^{19}F-spectra of $F_2P(S)$–O–$P(S)F_2$ can be interpreted in the same way; the normally appearing doublet of lines splits into two symmetrical nonets at very slow scan speeds (*111*).

4.4. Mössbauer investigations

The Mössbauer measurements are confined to a few halophosphates of tin and iron only; the data are listed in Table 6. The information to be taken from the isomer shift and the quadrupole splitting, is of importance in the discussion of bonding as well as for the consideration of structural problems. Occasionally the results of Mössbauer spectroscopy can be used supplementary to those of other spectroscopical methods for supplying indirect arguments in the structural discussion of compounds which do not contain "Mössbauer nuclei". By the interpretation of the ^{119}Sn-Mössbauer spectra along with the vibrational spectra, the structure of the polymeric trimethyltin dichlorophosphate could be recognised as that of a five-coordinated tin compound with bridging dichlorophosphate groups of type (IVb), and in analogy the corresponding structure of the lead compound $[(CH_3)_3Pb(PO_2Cl_2)]_\infty$ could be determined from the vibrational spectra alone (*86*).

Principally, structural information can be obtained from the quadrupol splitting in the Mössbauer spectra of tin(IV)-compounds by the calculation of expectation regions

Table 6. Mössbauer spectra of halophosphates of ^{119}tin and ^{57}iron [a]

Compound	δ_{IS} [mm/sec]	Δ [mm/sec]	Γ_1 [mm/sec]	Γ_2 [mm/sec]	Ref.
$[(CH_3)_2Sn(PO_3F]_\infty$	1,38	4,16	1,20	1,24	(*32*)
$[(CH_3)_2Sn(PO_2F_2)_2]_\infty$	1,53	5,13	0,90	0,92	(*43*)
$[C_2H_5)_2Sn(PO_2F_2)_2]_\infty$	1,67	4,91	1,10	1,10	(*43*)
$[(nC_3H_7)_2Sn(PO_2F_2)_2]_\infty$	1,64	5	0,97	0,95	(*43*)
$[(nC_4H_9)_2Sn(PO_2F_2)_2]_\infty$	1,67	5,03	1,04	1,06	(*43*)
$[(nC_8H_{17})_2Sn(PO_2F_2)_2]_\infty$	1,62	4,79	1,01	1,26	(*43*)
$[Cl_3Sn(PO_2Cl_2)\cdot POCl_3]_2$	1,32 (298 K)	0,30	1,33		(*112*)
	1,25 (77 K)	0,75	1,40		
$[(CH_3)_3Sn(PO_2Cl_2)]_\infty$	1,26	4,01	0,70	0,72	(*86*)
$[(CH_3)_2Sn(PO_2Cl_2)_2]_\infty$	1,39	4,66	0,96	0,93	(*86*)
$[Fe(PO_2F_2)_3]_\infty$	0,280 (300 K)	0,35	0,38		(*41*)
	0,325 (178 K)	0,35	0,38		
	0,345 (123 K)	0,36	0,38		
	0,360 (78 K)	0,35	0,37		
$[Fe(PO_2Cl_2)_3]_\infty$	0,285 (300 K)	0,17	0,29		(*41*)
	0,335 (178 K)	0,18	0,29		
	0,350 (123 K)	0,18	0,28		
	0,365 (78 K)	0,18	0,28		

[a] isomer shift δ_{IS} related to SnO_2 and $Na_2[Fe(CN)_5NO]\cdot 2\,H_2O$ respectively ΔE^Q = quadrupol splitting, Γ_i = half width

on the basis of point charge models (*113*) under the assumption of partial field gradients (*114*). According to this the alkyl groups of compounds $[R_2Sn(PO_2X_2)_2]_\infty$ (Table 6) with X = F, Cl are doubtlessly situated in trans position, with a symmetry D_{4h} at the tin atom (*43*). In the same way the five-coordination of the tin atoms in $[(CH_3)_2Sn(PO_2Cl_2)]_\infty$ can be seceared by the significantly smaller quadrupol splitting of ΔE^Q = 4.01 mm/sec. Also it seems possible, that the polymeric $[(CH_3)_2Sn(PO_3F)]$ is of the same structural type, because its quadrupol splitting (4.16 mm/sec) is of the same order of magnitude and gives rise to doubts with respect to a six-coordination (*32*). In the Mössbauer spectra of iron halophosphates recorded up to now there are remarkable differences between $Fe(PO_2Cl_2)_3$ and $Fe(PO_2F_2)_3$, as far as the quadrupol splitting and the line width are concerned (*41*). Both parameters are by far higher in the fluorophosphate, so that in this case the crystal field gradient gives a bigger contribution to the field gradient at the nucleus than in the chloro compound. This can be understood by assuming a statistical participation of FeF ligands besides the octahedral FeO coordination in $Fe(PO_2F_2)_3$, so that the iron atoms have a site symmetry differing from the point group O_h. On the other hand in $Fe(PO_2Cl_2)_3$ there can not possibly exist a coordination different from FeO_6; this is in good accordance with vibrational spectroscopy (*41*).

References

1. *Grunze, H.:* Z. Chem. *3*, 297 (1963).
2. *Schmutzler, R.:* Advances in Fluorine Chem. *5*, 31 (1965).
3. *Lange, W., Livingston, R.:* J. Amer. Chem. Soc. *72*, 1280 (1950).
4. *Bernstein, P.A., Hohorst, F.A., Eisenberg, M., Desmarteau, D.D.:* Inorg. Chem. *10*, 1549 (1971).
5. *Topchiev, A. V., Andrawo, V.N.:* Dokl. Akad. Nauk SSSR *111*, 365 (1956).
6. *Grunze, H.:* Z. anorg. allg. Chem. *298*, 152 (1959).
7. *Falius, H., Giesen, K.P.:* Angew. Chem. *83*, 578 (1971): Internat. Ed. *10*, 555 (1971).
8. a) *Grunze, H., Wolf, G.-U.:* Z. Chem. *3*, 354 (1963);
   b) – – Z. anorg. allg. Chem. *329*, 56 (1964).
9. *Lange, W., Askitopoulos, K.:* Ber. dtsch. Chem. Ges. *71*, 801 (1938).
10. *Grunze, H., Meisel, M.:* Z. Naturforsch. *18b*, 662 (1963).
11. *Fluck, E., Steck, W.:* Z. anorg. allg. Chem. *388*, 53 (1972).
12. *Moisier, L.C., White, W.E.:* Ind. Eng. Chem. *43*, 246 (1951).
13. *Chackalackal, S.M., Stafford, F.E.:* J. Amer. Chem. Soc. *88*, 4823 (1966).
14. *Cox, B.:* J. Chem. Soc. *1956*, 876.
15. *Lange, W., Livingston, R.:* J. Amer. Chem. Soc. *69*, 1073 (1947).
16. *Singh, E.B., Sinha, P.C.:* J. Indian Chem. Soc. *43*, 750 (1966).
17. *Neels, J., Grunze, H.:* Z. anorg. allg. Chem. *360*, 284 (1968).
18. *Grunze, H.:* Z. anorg. allg. Chem. *324*, 1 (1963).
19. *Fluck, E., Steck, W.:* Synthesis in Inorg. and Metal-org. Chem. *1*, 145 (1971).
20. *Goubeau, J., Schulz, P.:* Z. physik. Chem. *14*, 49 (1958).
21. a) *Falius, H.:* Angew. Chem. *82*, 702 (1970).
    b) – Internat. Ed. *9*, 733 (1970).
22. *Roesky, H.W.:* Chem. Ber. *100*, 1447 (1967).
23. *Scherp, A.:* Umschau *72*, 65 (1972).
24. *Lange, W.:* Ber. dtsch. Chem. Ges. *62*, 793 (1929).
25. *Singh, E.B., Sinha, P.C.:* J. Indian Chem. Soc. *41*, 407 (1964).
26. *Perloff, A.:* Acta cryst. *B28*, 2183 (1972).
27. *Atoji, M., Rundle, R.E.:* J. Chem. Phys. *29*, 1306 (1958).
28. *Battelle, L.F., Trueblood, K.N.:* Acta cryst. *19*, 531 (1965).
29. *Hill, O.F., Audrieth, L.F.:* J. Phys. a. Colloid Chem. *54*, 690 (1950).
30. *Hill, O.F., Audrieth, L.F.:* Inorg. Synth. III, 106 (1950).
31. *White, W.E., Munn, J.M., Gilliland, J.E., Wright, B.B.:* Ger. Offen. 1.961.583 (Ozark-Mahoning); Chem. Abstr. *75*, 51046a (1971).
32. *Chivers, T., Van Roode, J.H.G., Ruddick, J.N.R., Sams, J.R.:* Canad. J. Chem. *51*, 3702 (1973).
33. *Berndt, A.F.:* Acta cryst. *30B*, 529 (1974).
34. *Lange, W.:* Ber. dtsch. Chem. Ges. *62*, 786 (1929).
35. *Harrison, R.W., Thompson, R.C., Trotter, J.:* J. Chem. Soc. (A) *1966*, 1775.
36. *Wannagat, V., Rademachers, J.:* Z. anorg. allg. Chem. *289*, 66 (1957).
37. a) *Kreshkov, A.P., Drozdov, V.A., Orlova, I. Yu.:* Zh. Abshch. Khim. *36*, 525 (1966);
    b) – – – Chem. Abstr. *65*, 735h (1966).
38. *Roesky, H.W.:* Chem. Ber. *100*, 2147 (1967).
39. *Schaible, B., Haubold, W., Weidlein, J.:* Z. anorg. allg. Chem. *403*, 289 (1974).
40. *Weidlein, J.:* Z. anorg. allg. Chem. *358*, 13 (1968).
41. *Pebler, J., Dehnicke, K.:* Z. Naturforsch. *26b*, 747 (1971).
42. *Roesky, H.W., Wiezer, H.:* Chem. Ber. *104*, 2258 (1971).
43. *Tan, T.H., Dalziel, J.R., Yeats, P.A., Sams, J.R., Thompson, R.C., Aubke, F.:* Canad. J. Chem. *50*, 1843 (1972).

44. *Eisenberg, M., Desmarteau, D.D.:* Inorg. Chem. *11*, 1901 (1972).
45. *Thompson, R.C., Reed, W.:* Inorg. nucl. Chem. Letters *5*, 581 (1969).
46. *Babaeva, V.P., Rosolovskii, V. Ya.:* Russ. J. Inorg. Chem. *16*, 471 (1971).
47. *Grunze, H., Jost, K.H., Wolf, G. U.:* Z. anorg. allg. Chem. *365*, 294 (1969).
48. a) *Appel, H.R.:* U.S. Patent 2842605 (1968).
    b) – Chem. Abstr. *52*, 19107g (1958).
49. *Biermann, U., Glemser, O.:* Chem. Ber. *102*, 3342 (1969).
50. *Dalziel, J.R., Klett, R.D., Yeats, P.A., Aubke, F.:* Canad. J. Chem. *52*, 231 (1974).
51. *Kreshkov, A.P., Drozdov, V.A., Orlova, I. Yu.:* Zh. Obshch. Khim. *36*, 2014 (1966).
52. *Glemser, O., Biermann, U., von Halasz, S.P.:* Inorg. nucl. Chem. Letters, *5*, 501 (1969).
53. a) *Ryss, I.G., Tul'chinskii, V.B.:* Zh. Neorgan. Khim. *7*, 1313 (1962);
    b) – – Chem. Abstr. *57*, 6852c (1962).
54. a) *Tul'chinskii, V.B., Ryss, I.G., Zubov. V.I.:* Zh. Neorgan. Khim. *11*, 2694 (1966);
    b) – – – Chem. Abstr. *66*, 51755v (1967).
55. *Lustig, M., Ruff, J.K.:* Inorg. Chem. *6*, 2115 (1967).
56. *Siebert, H.:* Anwendungen der Schwingungsspektroskopie in der anorganischen Chemie, Springer, Berlin–Heidelberg–New York 1966, S. 70.
57. *Stoelzer, C., Simon, A.:* Z. anorg. allg. Chem. *339*, 38 (1965).
58. *Cavell, R.G., Leary, R.D., Tomlinson, A.J.:* Inorg. Chem. *11*, 2573 (1972).
59. *Roesky, H.W., Dietl, M., Norbury, A.H.:* Z. Naturforsch. *28b*, 707 (1973).
60. *Colburn, C.B., Hill, W.E., Sharp, D.W.A.:* J. Chem. Soc. (A) *1970*, 2221.
61. *Robinson, E.A.:* Canad. J. Chem. *40*, 1725 (1962).
62. *Binder, H., Fluck, E.:* Z. anorg. allg. Chem. *365*, 170 (1969).
63. *Schmidt, M., Schmidbaur, H., Binger, A.:* Chem. Ber. *93*, 872 (1960).
64. *Grunze, H., Meisel, M.:* Z. Chem. *9*, 346 (1969).
65. *Dehnicke, K., Schmitt, R.:* Z. anorg. allg. Chem. *358*, 1 (1968).
66. *Goubeau, J., Schulz, P.:* Z. anorg. allg. Chem. *294*, 224 (1958).
67. *Kauffmann, S., Dehnicke, K.:* Z. anorg. allg. Chem. *347*, 318 (1966).
68. *Klopsch, A., Dehnicke, K.:* Z. Naturforsch. *27b*, 1304 (1972).
69. *Shihada, A.-F., Dehnicke, K.:* Z. Naturforsch. *28b*, 268 (1973).
70. *Schaible, B., Weidlein, J.:* Z. anorg. allg. Chem. *403*, 301 (1974).
71. *Moras, D., Mitschler, A., Weiss, R.:* Acta Cryst. *B25*, 1720 (1969).
72. *Weidlein, J., Dehnicke, K.:* Chimia *27*, 375 (1973).
73. *Nyborg, J., Danielsen, J.:* Acta Chem. Scand. *24*, 59 (1970).
74. *Grunze, H.:* Z. Chem. *6*, 266 (1966).
75. *Müller, H., Dehnicke, K.:* Z. anorg. allg. Chem. *350*, 231 (1967).
76. *Meisel, M., Grunze, H.:* Z. anorg. allg. Chem. *400*, 128 (1973).
77. *Danielsen, J., Rasmussen, S.E.:* Acta Chem. Scand. *17*, 1971 (1963).
78. *Saavedra, A.:* Dissertation Universität Marburg (in Vorbereitung).
79. *Schimitschek, E.J., Trias, J.A.:* Inorg. nucl. Chem. Letters, *6*, 761 (1970).
80. *Dehnicke, K., Shihada, A.-F.:* Z. Naturforsch. *28b*, 148 (1973).
81. a) *Bassett, H., Taylor, H.S.:* J. Chem. Soc. *99*, 1402 (1911);
    b) – – Z. anorg. allg. Chem. *73*, 75 (1912).
82. *Schmidt, M., Schmidbauer, H., Ruidisch, I.:* Angew. Chem. *73*, 408 (1961).
83. *Dehnicke, K.:* Z. anorg. allg. Chem. *308*, 72 (1961).
84. *Dehnicke, K.:* Z. anorg. allg. Chem. *309*, 266 (1961).
85. *Dehnicke, K.:* Chem. Ber. *98*, 290 (1965).
86. *Dehnicke, K., Schmitt, R., Shihada, A.-F., Pebler, J.:* Z. anorg. allg. Chem. *404*, 249 (1974).
87. *Lindquist, I., Zackrisson, M., Erikson, S.:* Acta Chem. Scand. *13*, 1758 (1959).
88. *Fluck, E.:* Angew. Chem. *72*, 752 (1960).
89. *Grunze, H., Jost, K.H.:* Z. Naturforsch. *20b*, 268 (1965).
90. *Trotter, J., Whitlow, S.H.:* J. Chem. Soc. (A) *1967*, 1383.

91. *Pauling, L.:* The Nature of the Chemical Bond, Cornell University Press, Ithaca, 3rd Edn., 1960, Ch. 7.
92. *Schomaker, V., Stevenson, D.P.:* J. Amer. Chem. Soc. *63*, 37 (1941).
93. *Robinson, E.A.:* Canad. J. Chem. *41*, 3021 (1963).
94. *Harrison, R. W., Trotter, J.:* J. Chem. Soc. (A) *1969*, 1783.
95. *Cruickshand, D. W. J., Robinson, E.A.:* Spectrochim. Acta *22*, 555 (1966).
96. *Bühler, K., Bues, W.:* Z. anorg. allg. Chem. *308*, 62 (1961).
97. *Durand, J., Granier, W., Vilminot, S., Cot, L.:* C.R. Acad. Sci. *C275*, 737 (1972).
98. *Durand, J., Granier, W., Cot, L.:* C.R. Acad. Sci. *C277*, 101 (1973).
99. *Durand, J., Granier, W., Cot, L., Avineus, C.:* C. R. Acad. Sci. *C277*, 13 (1973).
100. *Galigue, J. L., Durand, J., Cot, L.:* Acta Cryst. *B30*, 697 (1974).
101. *Walford, L. K.:* Acta Cryst. *22*, 324 (1967).
102. *Schimitschek, E. J., Trias, J. A., Liang, C. Y.:* Spectrochim. Acta *27A*, 2141 (1971).
103. *Wartenberg, E. W., Goubeau, J.:* Z. anorg. allg. Chem. *329*, 269 (1964).
104. *Schrader, B., Meier, W.:* DMS Raman/IR-Atlas organischer Verbindungen, Verlag Chemie Weinheim 1974, Nr. B4-01.
105. *Corbridge, D. E. C., Lowe, E. J.:* J. Chem. Soc, *1954*, 4555.
106. *Delwaulle, M.-L., Francois, F.:* C. R. Acad. Sci. *226*, 894 (1948).
107. *Reddy, G. S., Schmutzler, R.:* Z. Naturforsch. *20b*, 104 (1965).
108. *Reddy, G. S., Schmutzler, R.:* Z. Naturforsch. *25b*, 1199 (1970).
109. *Fluck, E.:* Chem. Ber. *94*, 1388 (1961).
110. *Moedritzer, K., Maier, L., Groenweghe, L. C. D.:* J. chem. Engng. Data 7, 307 (1962).
111. *Hill, W. E., Sharp, D. W., Colburn, C. B.:* J. Chem. Phys. *50*, 612 (1969).
112. *Moras, D., Weiss, R.:* Acta Cryst. *B25*, 1726 (1969).
113. a) *Parish, R. V., Platt, R. H.:* Inorg. chim. Acta *4*, 65 (1970);
b) *Clark, M. G., Maddock, A. G., Platt, R. H.:* J. Chem. Soc. Dalton *1972*, 281;
c) *Maddock, A. G., Platt, R. H.:* J. Chem. Soc. (A) *1971*, 1191.
114. *Lorbert, J., Pebler, J., Lange, G.:* J. Organometal. Chem. *54*, 177 (1973).
115. *van Wazer, R., Fluck, E.:* J. Amer. Chem. Soc. *81*, 6360 (1959).
116. *Viard, B., Amaudrut, J., Devin, C.:* Bull. Soc. Chim. France *1975*, 1940.
117. *Brown, S. D., Emme, L. M., Gard, G. L.:* J. inorg. nucl. Chem. *37*, 2557 (1975).
118. *Eberwein, B.:* Dissertation Universität Stuttgart 1975.

Note added in proof:
Recently the halophosphates $[TiOCl(PO_2Cl_2) \cdot POCl_3]_\infty$ (*116*), $Cr(PO_2F_2)_3$, $K_2[MO_2(PO_2F_2)_4]$ (M = Cr, Mo, W) (*117*) and several tetramethylstibonium dihalophosphates (*118*) have been prepared and characterised by vibrational spectroscopy.

Dithiocarbamates of Transition Group Elements in "Unusual" Oxidation States

J. Willemse, J.A. Cras, J.J. Steggerda

Department of Inorganic Chemistry, University of Nijmegen, Toernooiveld, Nijmegen, The Netherlands

C.P. Keijzers

Department of Physical Chemistre, University of Nijmegen, Toernooiveld, Nijmegen, The Netherlands

Table of Contents

J. Willemse, J. A. Cras, J. J. Steggerda and C. P. Keijzers

Introduction

The oxidation state concept

The concept of the oxidation state of an element in a chemical compound has a long and confusing history. In the most pretentious form of the concept the oxidation state or oxidation number is the electrical charge localised on the concerned atom in the compound. Confusion arises when we realise that the definition of the atomic domain is arbitrary and the experimental determination of the electrical charge in the once chosen domain is often problematic.

The so-called 'formal' oxidation number is the widest known operational concept for oxidation state. This number can be determined from the stoichiometry of the compound by some simple rules (the oxidation number of O and H being − 2 and + 1, respectively; the sum of the oxidation numbers is equal to the total charge of the species under consideration). The statement that the formal oxidation number of Cu in $CuCl_2$ is + 2 was theoretically significant in a simple but obsolete bonding model in which $CuCl_2$ was thought to be composed of isolated Cu^{2+} and Cl^- ions.

There is now a well-founded scepticism about the formal oxidation number, as there is experimental and theoretical evidence that in many chemical compounds these formal oxidation numbers have little correlation with the charge distribution in the molecule. Notwithstanding this, the formal oxidation number still has great popularity because:

- the assignment is easy,
- on the basis of the formal oxidation state a great number of inorganic compounds can be classified in a sensible and practical way,
- the crystal field theory can give a good comprehension of some properties (magnetism, coordination geometry, etc.) of complexes of transition elements, there is a straightforward relation between d-electron configuration and formal oxidation number.

In this review the oxidation state or oxidation number will be used in this formal sense, unless otherwise stated.[1]

The unusual oxidation state

According to Nyholm and Tobe (*1*) "The definition of a usual oxidation state refers to oxidation states that are stable in environments made up of those chemical species that were common in classical inorganic compounds, e.g. oxides, water and other simple

[1] For a detailed discussion about the oxidation states the reader is referred to the well-known book of *C. K. Jørgensen:* Oxidation Numbers and Oxidation States. Berlin–Heidelberg–New York: Springer 1969.

oxygen donors, the halogens, excluding fluorine, and sulfur". For example, copper compounds with these ligands have the oxidation state + 1 and + 2, whereas in the fluorine compound K_3CuF_6 the formal oxidation number + 3 must be assigned to the transition metal. Only because of the high electronegativity of fluorine can copper reach this so-called unusual oxidation state.

The situation is more complicated with polyatomic ligands which have σ-donor as well as π-acceptor properties. The stability of complexes with unusually high or low oxidation states of the central metal atom depends on the possibility for charge levelling by σ-bonding and π-back bonding. If in $Co(NO)_3$ the NO is considered as a positively charged ligand, Co is in the formal oxidation state of -3, which is stabilised by the weak σ-donation and the strong π-backbonding of NO^+, due to its positive charge.

The Ni(0) compound $Ni(CO)_4$ is stable because the σ- and π-bonding nearly cancel their charge shifts. In $Ni(CN)_4^{2-}$ Ni(II) is surrounded by CN^- ions, which donate electrons via σ-bonding, the π-back-donation being small because of the negative charge of these ligands. This back-donation can be increased by the use of the very strong reducing agent metallic potassium, giving the compound $K_4Ni(CN)_4$ with Ni in the low oxidation state 0, which is fully comparable with $Ni(CO)_4$.

Among the polyatomic ligands the dithiocarbamato ligand can stabilise high oxidation states of the transition metals in its complexes. Like all 1,1-dithioates the σ-donation and π-back-donation of the sulfur atoms is assumed to be of the same order of magnitude.

The special feature of the dithiocarbamato ligand is an additional π-electron flow from the nitrogen atom to the sulfur atoms via a planar delocalised π-orbital system. The net effect is a strong electron donation, resulting in a high electron density on the metal.[2])

Infrared evidence supports the suggestion that the lone pair of the nitrogen atom in the dithiocarbamato complex becomes progressively more important for the donation of electrons the higher the oxidation state of the metal.

So the stability of compounds with the metal in high oxidation numbers is thought to be possible because this formal oxidation number in fact does not reflect the real

2 The reverse effect is found in the dithiophosphinato complexes, where an additional π-electron flow from the sulfur to the empty orbitals of the phosphorus is present, causing stabilisation of a low oxidation state of the transition metal.

charge on the metal: the difference between the formal and real charge being greater here than in other systems.

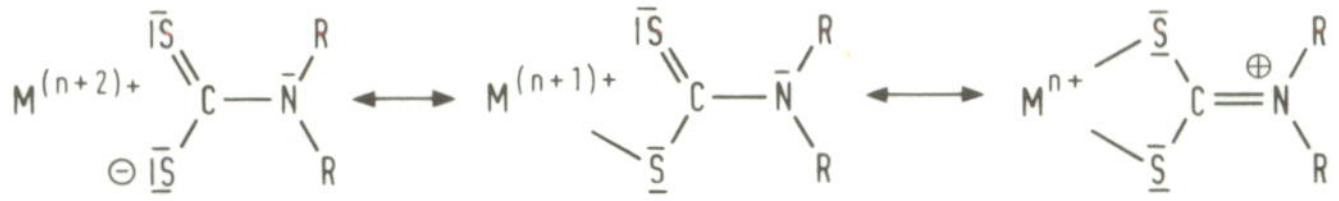

In view of this, it is not surprising that dithiocarbamato compounds with copper in the oxidation state + 3 are stable; instead it must be regarded as unexpected that Cu(I) dithiocarbamato complexes exist. The latter complexes are not simply monomeric, but they are tetrameric metal cluster compounds. Obviously, the stability must be attributed to the metal-metal bond rather than to the stabilising effect of the ligand.

The same holds for the hexameric Ag(I) and dimeric Au(I) dithiocarbamates. In all other dithiocarbamato complexes in which the metal has a low oxidation state the existence of this type of compounds is due to other, low-oxidation-number stabilising ligands e.g. NO^+ in $(NO)_2Fe(Et_2dtc)_2$ and CO in $(CO)_4Fe(Et_2dtc)$.

The ambiguity of oxidation numbers

In complexes of the dithiolene type ligands the oxidation number of the metal cannot unambiguously be determined, due to the uncertainty of the charge on the ligand. $R_2C_2S_2$ can be considered either as a dinegative dithiolato anion or as a neutral dithioketo ligand.

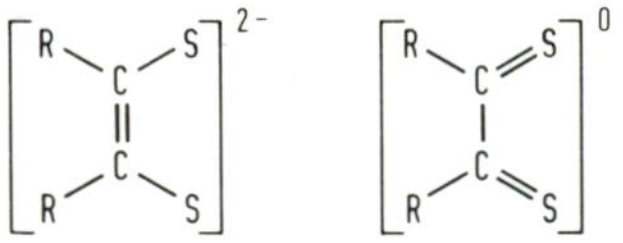

In dithiocarbamato complexes such an ambiguity can only occur when at least two dithiocarbamato ligands are bonded to a metal. In that case the question arises whether the compound is a bis(dithiocarbamato) or a thiuram disulfide complex. In these two types of complexes the oxidation number of the metal differs 2 units.

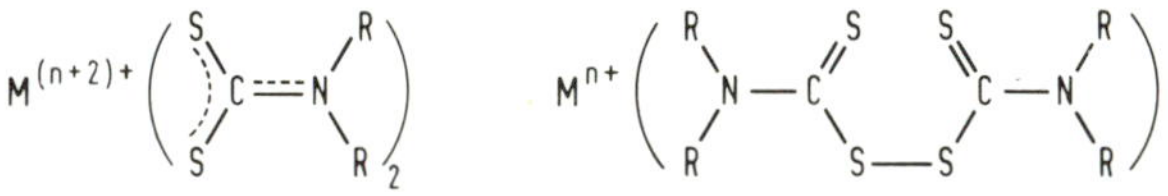

This ambiguity, however, can be overcome rather easily by X-ray analysis and sometimes by infrared or ESCA spectroscopy.

Another ambiguity may arise when two or more atoms of an element occur at unequal sites in the compound (e.g. $Na_2S_2O_3$, AgO), this is often found in the dithiocarbamato compounds of group Ib metals, e.g. $Cu_3(Bu_2\textit{dtc})_6(Cd_2I_6)$. The unequality can be detected by a full X-ray structure determination but sometimes also by more simple spectroscopic or magnetic measurements. How oxidation numbers must be attached to such unequal sites is often rather arbitrary, although spectroscopic and magnetic data as well as comparison with related substances can be important guides.

With a few exceptions, only the binary dithiocarbamato complexes of the transition metals and those of group IIb are included, leaving it to the reader, to determine what has to be considered as usual or unusual.

For a more complete survey of the dithiocarbamato chemistry up to 1969 the reader is referred to the reviews of *Coucouvanis* (*2*) and *Eisenberg* (*2a*). The investigations about the fluxionality of the octahedral dithiocarbamato complexes are not covered in this article, as they were reviewed recently by *Pignolet* (*3*).

J. Willemse, J.A. Cras, J.J. Steggerda and C.P. Keijzers

Group IV Transition Metal Dithiocarbamato Complexes

Titanium, Zirconium, Hafnium

Titanium and zirconium dithiocarbamates are prepared with the metal in the oxidation state + 4, whereas the hafnium dithiocarbamato chemistry is unknown.

All *Ti(IV)* complexes can be formulated as $Ti(R_2dtc)_nX_{4-n}$ with $n = 4, 3, 2$.

$Ti(R_2dtc)_4$ was prepared starting from $Ti(NR_2)_4$ (*4, 5*) or $TiCl_4$ (*6*). The structure of the ethyl compound was elucidated by X-ray studies (*7, 8*). The titanium atom is eight-coordinated by the sulfur atoms of the four chelating ligands and the coordination geometry closely approximates to that of a dodecahedron.

$Ti(R_2dtc)_3Cl$ is a seven coordinated monomeric species, as is suggested from the spectral data and the non-electrolytic character and confirmed by an X-ray diffraction study (*6*). Apart from the chlorine compounds the bromine analogues are reported (*9*).

$Ti(R_2dtc)_2Cl_2$ complexes are, in view of the spectral data, the conductance studies and the dipole measurements in benzene, six-coordinated cis compounds (*9*).

The *Zr(IV)* dithiocarbamato chemistry is restricted to the complexes $Zr(R_2dtc)_4$ until now (*4, 5*). The properties of these compounds are similar to those of $Ti(R_2dtc)_4$. Although no X-ray studies are performed, it is likely that these compounds are isostructural.

Group V Transition Metal Dithiocarbamato Complexes

Vanadium

Vanadium dithiocarbamates are reported with the metal in the oxidation states + 3 and + 4.

The *V(III)* containing complexes $V(R_2dtc)_3$ were prepared for R = Et (*5*), R = Bz and R_2 = MePh (*10*). The magnetic moments of these air-sensitive compounds are characteristic of two unpaired electrons as is expected for a d^2 configuration, the coordination geometry is assumed to be a distorted octahedron (D_3).

V(IV) complexes with the formula $V(R_2dtc)_4$ (R = Me, Et) were studied by *Bradley et al.* (*11, 12, 4, 5*). The ethyl complex (*5*) is thermally unstable and air-sensitive. The thermal instability accounts for the formation of the vanadium tris(dithiocarbamate). The tetrakis(dithiocarbamato) complex is isomorphous with $Ti(Et_2dtc)_4$ (*4*) and contains dodecahedrally coordinated vanadium. This is supported by the EPR and electronic spectra, which could be interpreted reasonably in terms of crystal field theory (*5*).

Niobium

The oxidation states of the metal in niobium dithiocarbamates can be + 4 and + 5.

The *Nb(IV)* dithiocarbamato complex $Nb(R_2dtc)_4$ was prepared either by the reaction of CS_2 with $Nb(NR_2)_5$ (R = Me, Et (*4*)), in which the metal is reduced, or by the reaction of $NbCl_4$ with sodium or ammonium dialkyldithiocarbamate (R = Et (*13, 14*), $R_2 = (CH_2)_4$ (*8*)). Whereas all authors assume the metal to be eight-coordinated, there is no unanimity about the magnetic moment of the complexes. Values reported are 0.5 BM (*4, 13*) and 1.57 BM (*14, 15*) at room temperature.

An air-stable, deep-red coloured solid, with composition $Nb_2Br_3(Et_2dtc)_5$ is claimed to be the reaction product of $NbBr_4$ and $Na(Et_2dtc)$ in a 1 : 2 ratio (*13*). The infrared spectra suggest bidentate dithiocarbamate coordination, the magnetic susceptibility is markedly field-dependent. Although the compound is monomeric in $CHCl_3$, a molar conductance in nitromethane is found, which increases sharply on dilution, attributed to a weak electrolytic character. Accordingly, the authors formulate this compound as $[Nb_2(Et_2dtc)_5Br_2]Br$.

All known *Nb(V)* dithiocarbamato complexes are prepared by the reaction of NbX_5 and $Na(Et_2dtc)$, the nature and the number of products being dependent on the stoichiometry of the reacting species, the solvent, and the temperature (*16, 17*). No $Nb(Et_2dtc)_5$ could be synthesized, which is thought to be due to steric strain which liberates a dithiocarbamato radical. The isolated complexes can be formulated as $Nb(Et_2dtc)_4X$, $Nb(Et_2dtc)_3Cl_2$ and $Nb(Et_2dtc)_2X_3$ (X = Cl, Br).

$Nb(Et_2dtc)_4X$ is a 1 : 1 electrolyte in which the metal is probably eight-coordinated by sulfur atoms from four bidentate dithiocarbamate groups. Infrared data suggest that the ligands are unsymmetrically bonded.

On account of the 1 : 1 electrolytic character it is likely that $Nb(Et_2dtc)_2X_3$ is of the type $[Nb(Et_2dtc)_4^+]$ $[NbX_6^-]$. This is supported by the possibility of preparing these complexes from $Nb(Et_2dtc)_4X$ and $[R_4N]$ $[NbX_6]$. The molar weight determination, however, does not fit in this proposal, in CH_2Cl_2 a value of 757 is found for the chlorine species, whereas the molar weight should be 495 if the complex is fully dissociated in this solvent and 990 if no ionic dissociation occurs.

As a possible structure for $Nb(Et_2dtc)_3Cl_2$, an ionic dimer $[Nb_2(Et_2dtc)_6Cl_3]Cl$ is suggested in which two $Nb(Et_2dtc)_3$ units are bridged by three chlorine atoms. Two C–S absorption vibrations in the 1000 cm^{-1} region point either to an asymmetric coordination of the dithiocarbamates or to a monodentate binding of these ligands. Apart from the equivalent conductance and the molar weight (approximating the value required for monomeric $Nb(Et_2dtc)_3Cl_2$), no further data about this interesting compound are known.

Tantalum

The only tantalum dithiocarbamates known so far have the metal in the oxidation state + 5.

The major difference in the dithiocarbamato chemistry of tantalum and niobium is the existence of the *Ta(V)* species $Ta(R_2dtc)_5$, whereas the comparable niobium compound is unknown. The preparative methods for $Ta(R_2dtc)_5$ involve either a CS_2 insertion in $Ta(NR_2)_5$ (R = Me (*11*)) or a reaction between $TaCl_5$ and sodium or ammonium dithiocarbamate (R = Et (*7*), $R_2 = (CH_2)_4$ (*15*)). The structure of this interesting diamagnetic compound is not elucidated, but from infrared data the presence of one or more unidentate ligands is suggested (*11, 15*).

Heckley and coworkers reported the reaction of $TaCl_5$ and $Na(Et_2dtc)$ to give complexes analogous to those obtained in the niobium reactions (*16, 17*). The properties of $Ta(Et_2dtc)_4X$ and $Ta(Et_2dtc)_2X_3$ are analogous to those of the comparable niobium compounds.

Group VI Transition Metal Dithiocarbamato Complexes

Chromium

Chromium dithiocarbamates are known with the metal in the oxidation states + 2 and + 3.

The *Cr(II)* complexes $Cr(R_2dtc)_2$ (R = Me, Et and $R_2 = C_4H_8$) are very sensitive to oxidation and pyrophoric in air (*18*). They are prepared from $CrCl_2$ and $Na(R_2dtc)$ in aqueous or 50%-alcoholic solutions. X-ray powder diffraction studies revealed them to be isomorphous with the corresponding Mn(II), Fe(II), Zn(II), and Cu(II) compounds, and hence probably five-coordinated in dimers with bridging S atoms. Electronic reflection spectra, however, could be interpreted assuming a tetragonal structure. Magnetic measurements indicate antiferromagnetic coupling between the electron spins of both Cr atoms of the dimer (*19*).

When the synthesis is performed in pure alcohol, $Cr(R_2dtc)Cl$ is formed instead of $Cr(R_2dtc)_2$. The former compound, however, is not further characterised (*19*).

The *Cr(III)* compounds with the formula $Cr(R_2dtc)_3$ are easily prepared for various R groups. They are stable and well-studied (*2*). Electronic spectra (*20, 21*) and EPR data (*22, 23*) are in accordance with an octahedral coordination of six S atoms around the metal. Voltametric data for the reduction and oxidation of $Cr(R_2dtc)_3$ have been published (*162*).

Interesting compounds are $[Cr(R_4tds)Cl_2]Cl$ (R = Me, Et) (*24*). Electronic and infrared spectra indicate that these complexes are not Cr(V) dithiocarbamato complexes but rather Cr(III) compounds with coordinated thiuram disulfide. As will be shown, thiuram disulfide can oxidise Cu, Ag and Au to M(II) and M(III) dithiocarbamato complexes. The Cr(III)-thiuram disulfide combination seems to be stable, just like the thiuram disulfide combination with Zn, Cd, and Hg.

Molybdenum

Molybdenum dithiocarbamates are reported with the metal in the oxidation states + 2, + 3, + 4, and + 5.

A *Mo(II)* compound of composition $Mo(R_2dtc)_2$ is claimed to exist (*25*). Upon standing, this complex is transformed into a product containing two sulfido bridges between two Mo atoms, to each of which a dithiocarbamato and a thiocarboxamido group ($R_2NC = S$) is coordinated (*25, 26*). It is interesting to see that when R_2dtc comes in contact with an electron-rich metal, an oxidative addition occurs in which R_2dtc is split into a thiocarboxamido and a sulfido group. Further examples are found in Rh chemistry (*27*).

Mo(III) complexes are reported with the formula $[Mo(R_2dtc)_3]_2$ (*28, 29*). In spite of considerable effort, we did not succeed in preparing these compounds by the method of Brown, $Mo(R_2dtc)_4$ always being the product, and we doubt whether tris complexes can be obtained by this method.

The preparation and properties of the *Mo(IV)* complexes $Mo(R_2dtc)_4$, with a variety of R groups have been studied (*15, 30, 31, 32, 33, 34*). Several preparative routes are described, but the oxidative decarbonylation of molybdenum carbonyls with thiuram disulfide is preferred, due to the high yields and purity of the products (*34*). The structure of $Mo(Et_2dtc)_4$ was elucidated by X-ray analysis (*35*). Mo is coordinated to four bidentate dithiocarbamato ligands, the MoS_8 coordination symmetry being that of a slightly distorted triangular dodecahedron, see Fig. 1 and 2. Comparing this structure with that of $W(Et_2dtc)_4Br$ (*36*), it is seen that the position of the sulfur atoms is approximately equal but the position of the ligands is different. In the tungsten compound the ligands are located in the *xz* and *yz* planes, whereas in $Mo(Et_2dtc)_4$ each ligand connects one sulfur atom in the *xz* plane with one sulfur atom in the *yz* plane. The electronic spectra and the diamagnetism of the molybdenum compounds have been interpreted in crystal field and Extended Hückel MO terms (*37*).

A recent report mentions the preparation of $Mo(R_2dtc)_2X_2$ (R = Et, $R_2 = (CH_2)_5$; X = Cl, Br) from a benzene suspension of $OMo(R_2dtc)_2$ and gaseous hydrogen halide (*38*). Unfortunately no properties of the resulting compound are given.

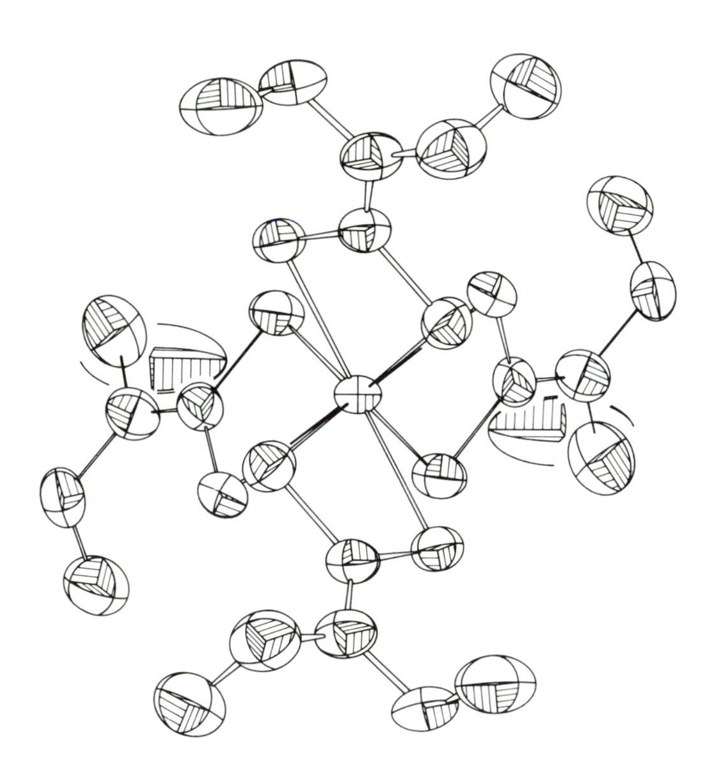

Fig. 1. The crystal structure of $Mo(Et_2dtc)_4$.

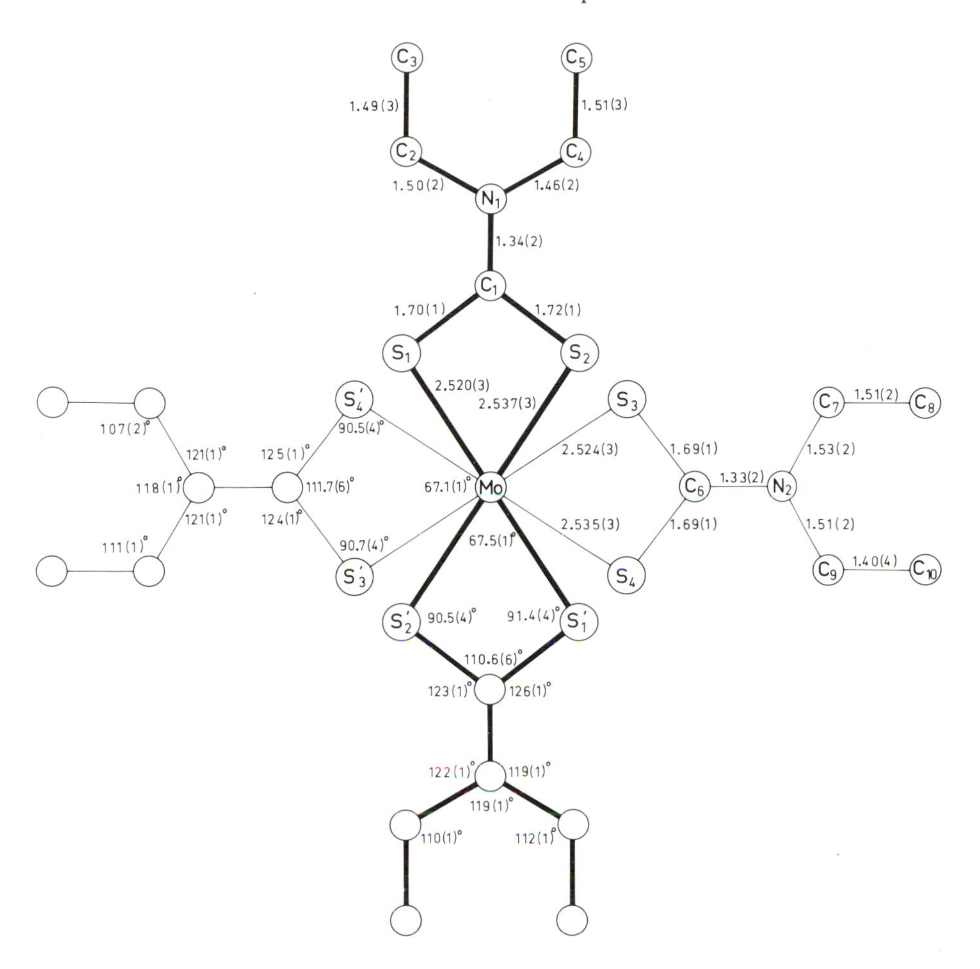

Fig. 2. Bond angles (degrees) and distances (Å) with e.s.d.'s of the $Mo(Et_2dtc)_4$.

Likewise without further detail, Chatt and coworkers mention the existence of compounds $XMo(R_2dtc)_3$ (X = Cl, Br, NCS) as somewhat air-sensitive crystalline solids (*39*).

The *Mo(V)* compounds $[Mo(R_2dtc)_4^+]X^-$ can be prepared by oxidative addition of thiuram disulfide to molybdenum carbonyls or by mild oxidation of $Mo(R_2dtc)_4$. In particular, the oxidation with tetraethylthiuram disulfide is interesting, as it yields the compound $[Mo(Et_2dtc)_4^+]$ $[Et_2dtc^-]$. They are stable, paramagnetic, ionic compounds (*32, 33*) in which molybdenum is eight-coordinated. Spectral data can be interpreted using a crystal field model of D_{2d} symmetry. The structures of these complexes are probably very similar to those of $[W(R_2dtc)_4^+]X^-$, the coordination geometry approximating a triangular dodecahedron.

Kirmse and *Hoyer* (*40*) reported the preparation of the paramagnetic 1 : 2 electrolyte $[Mo_2(Et_2dsc)_4Cl_4]Cl_2$. No sulfur analogue is yet known. The existence of this complex makes the formulation of the earlier-mentioned compound $[Cr(R_4tds)Cl_2]Cl$ remarkable.

Voltametric studies (*34, 37*) have revealed the electron transfers $Mo(R_2dtc)_4^{-1} \leftarrow Mo(R_2dtc)_4^{0} \rightleftharpoons Mo(R_2dtc)_4^{+1} \rightarrow Mo(R_2dtc)_4^{+2}$. The half wave potentials for the processes $0 \rightleftharpoons +1$ and $+1 \rightarrow +2$ depend upon the nature of R. The dependency can be described by the Taft relation $E_{1/2} = \rho \Sigma \sigma^*$ where σ^* is the Taft constant for the N-bonded substituents R. The rather low values of ρ indicate that the redox orbitals have mainly a metal character, the mixing of these orbitals with those of the ligands is rather small. This conclusion is in accordance with the interpretation of the electronic spectra and the results of Extended Hückel MO calculations (*37*). Only the compound with R = Ph does not fit into the Taft relation, the $E_{1/2}$ value is lower than expected, probably because of a mesomeric effect.

Tungsten

The oxidation states + 4 and + 5 are known for tungsten in its dithiocarbamato complexes.

The *W(IV)* compounds $W(Et_2dtc)_4$ (*14*) and $W[(CH_2)_4dtc]_4$ (*15*) are characterised only by analytical data and magnetic susceptibility measurements. They are considered to be paramagnetic with moments of about 1.0 BM, which is in contrast with the diamagnetism of $Mo(Et_2dtc)_4$. $W(R_2dtc)_4$ could not be prepared by the reaction of $W(CO)_6$ with R_4tds (*37*).

W(V) complexes with the formula $[W(R_2dtc)_4^+]X^-$ are described with various R and X. The structure of $[W(Et_2dtc)_4^+]Br^-$ was determined by X-ray analysis (*36*). Tungsten is coordinated to the eight sulfur atoms from the four bidentate dithiocarbamato ligands. The coordination geometry approximates very closely a triangular dodecahedron. Similar coordination has been found in $Ti(R_2dtc)_4$ (*8*) whereas a different eight-coordination geometry is found in $Th(R_2dtc)_4$ and $NEt_4Np(R_2dtc)_4$ (*41*). The preparation and the properties of $[W(R_2dtc)_4]X$ are very similar to those of the corresponding molybdenum compounds.

The electron transfer series $W(R_2dtc)_4^{0} \leftarrow W(R_2dtc)_4^{+1} \rightarrow W(R_2dtc)_4^{+2}$ was detected by voltametric measurements (*37*), the half wave potentials are 0.24 and 0.36 V lower than the corresponding values of the molybdenum compounds. Both transfers obey the Taft relation with a low value of ρ, which points to a rather small contribution of the ligand orbitals into the redox orbitals.

Group VII Transition Metal Dithiocarbamato Complexes

Manganese

Manganese dithiocarbamates are known with the metal in the oxidation state + 2, + 3 and + 4.

The *Mn(II)* complexes $Mn(R_2dtc)_2$ are very air sensitive and readily oxidised to $Mn(R_2dtc)_3$ (*42*).

Hendrickson, Martin and *Rohde* (*43*) reported the preparation of $Mn[(CH_2)_4dtc]_2$ and its electrochemical oxidation at a potential of + 0.48 V.

The reports about the crystal structure of $Mn(Et_2dtc)_2$, determined by means of X-ray powder diagrams are contradictory. According to *Fackler* and *Holah* (*18*) this compound is isomorphous with $Cu(Et_2dtc)_2$, but *Lahiry* and *Anand* (*44*) state the complex to be isostructural with $Ni(Et_2dtc)_2$. EPR data ($g_{\parallel} = 1.92$ and $g_{\perp} = 4.11$) and magnetic susceptibility measurements (4.1 BM at room temperature) show the compound to be the first Mn(II) complex with a quartet ground state (*44*).

A recent report of *Hill et al.* (*192*) states that the low values of the magnetic susceptibility of $Mn(Et_2dtc)_2$ obtained by *Lahiry* and *Anand* are due to partial oxidation of the sample and that the compound has a sextet ground state. The dimeric $[Mn(Et_2dtc)_2]_2$ is not X-ray isomorphous with $Ni(Et_2dtc)_2$, but gives a powder pattern similar to, though not identical with $Cu(Et_2dtc)_2$ according to these authors (*192*).

Apart from the bis(dithiocarbamato) complexes a tris(dithiocarbamato) complex of Mn(II), $Mn[(CH_2)_4dtc]_3^-$, is known. Although isolation of this type of complex failed so far (*45*), electrochemical characterisation was possible (*43*).

Stable *Mn(III)* compounds, $Mn(R_2dtc)_3$, have been known for a long time (*42, 46*). The structure of $Mn(Et_2dtc)_3$ is elucidated (*47*). The inner geometry of the $Mn(CS_2)_3$ core does not conform to the usual D_3 point symmetry of transition metal complexes of this type, but shows a strong distortion attributed to the Jahn-Teller effect. The electronic spectrum (*48, 49*) and the magnetic properties of this type of complexes are well studied (*50*).

Recently *Mn(IV)* containing complexes with $Mn(R_2dtc)_3^+$ cations were reported (*51, 52, 53, 54*). For $Mn(Et_2dtc)_3BF_4$ a magnetic moment of 3.74 BM is found (*52*), corresponding with the value expected for a high-spin d^3 Mn(IV) complex. The molecular structure of the core atoms of $Mn[(CH_2)_5dtc]_3ClO_4 \cdot CHCl_3$ approximates D_3 symmetry. The twist angle φ between the upper and the lower triangles, relative to the threefold axis (see Fig. 3) is 38° (being 60° and 0° for octahedron and trigonal prisms, respectively) (*55*).

The Mn–S distances vary from 2.32 (1) to 2.35 (1) Å, whereas in $Mn(Et_2dtc)_3$ the manganese atom is surrounded by three opposing pairs of sulfur atoms at distances of 2.38 (1), 2.43 (1) and 2.55 (1) Å (*47*). Electrochemical studies (*43*) in acetone and dichloromethane showed one-electron transfer series:

$$Mn(R_2dtc)_3^{+1} \rightleftharpoons Mn(R_2dtc)_3^{0} \rightleftharpoons Mn(R_2dtc)_3^{-1}$$

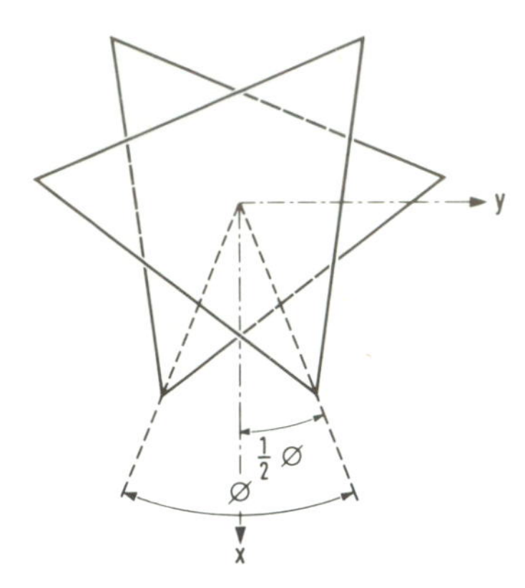

Fig. 3. Projection of the upper and the lower triangle of a geometry between an octahedron and a trigonal prism. Definition of the twist angle φ.

Martin c.s. (*43*) have pointed to the remarkable linear relation of the $E_{1/2}$ values of both processes, measured with a great variety of R. This indicates the relative abilities of various substituents to stabilise Mn(II) or Mn(IV) with respect to Mn(III); substituents which stabilise Mn(IV) destabilise Mn(II) and vice versa.

The influence of the N-bonded substituents R on the half-wave potentials can be described by a Taft relation, like is found for Mo, W and Au. The small value of ρ points to the dominance of metal orbitals in the redox orbital (*56*). The phenyl derivates do not fit this relation, probably because of a mesomeric influence. Here, however, the *n*-butyl and cyclohexyl also show small deviations, probably because of steric effects.

Technetium, Rhenium

No binary technetium dithiocarbamato complexes are reported.

Binary rhenium dithiocarbamato complexes are prepared with the metal in the oxidation state + 3, + 4 and + 5.

The simplest *Re (III)* dithiocarbamato compounds are $Re(R_2dtc)_3$. So far this type is only known for R = Et. It can be obtained by the reaction of $ReCl_3(MeCN)(PPh_3)_2$ and $Na(Et_2dtc)$. It is interesting to see that a similar reaction, using $Na(Ph_2dtc)$ produces $Re(Ph_2dtc)_3(PPh_3)$. Only the IR data from both compounds are published (*57*).

Another Re(III) complex, $Cl_2Re(R_2dtc)$, has a tetrahedral coordination around the metal (*58*).

At present only one *Re(IV)* dithiocarbamato complex has been described, a paramagnetic octahedral compound $Cl_2Re(Et_2dtc)$ (*59*). Obviously, the reported formula is in error.

Re(V) is found in the complexes with the cation $[Re(Et_2dtc)_4^+]$ which is synthesized with $[BPh_4^-]$ and the Re(I) species $[Re(CO)_3Cl(Et_2dtc)^-]$ as anions (*59*).
A polarographic study revealed one oxidation process to occur in acetone solution (*59*). The reduction waves reported have to be ascribed to the reduction of mercury dithiocarbamates, formed by a reaction of the electrode material (*60*).

Group VIII Transition Metal Dithiocarbamato Complexes

Iron

In the iron dithiocarbamates the metal can have the oxidation states + 2, + 3 and + 4.

Two types of *Fe(II)* dithiocarbamates are reported. To the first type belong the complexes $Fe(R_2dtc)_2$. The X-ray powder diffraction pattern shows the ethyl compound to be isomorphous with the five-coordinated, dimeric Cu(II) complex (*18*), so it is to be expected that the fifth Fe–S bond is longer than the other four. Further details about $Fe(R_2dtc)_2$ are scarce, as the compounds are air-sensitive and rapidly oxidise to Fe(III) complexes.

The other type consists of complexes $A^+[Fe(R_2dtc)_3^-]$ ($A^+ = R_4N^+, PPh_4^+$) in which the metal has the coordination number six (*61*). These compounds are very air-sensitive and their stability is dependent on the positive ion. Whereas with $[Et_4N^+]$ the complex is easily transformed to $Fe(R_2dtc)_2$, with $[PPh_4^+]$ the compound is more stable (*61*).

The most stable complexes in the iron dithiocarbamato chemistry are those with *Fe(III)*.

The compounds $Fe(R_2dtc)_3$ have a D_3 symmetry, with a twist angle φ varying from about 33 to 40°, dependending on the alkyl groups (*62, 63*).

The compounds are near the $^6A_1-^2T_2$ electronic cross over, subtle changes in substituents R being sufficient to change the position of the spin-state equilibrium (*63, 164*).

Recently *Chant c.s.* have published (*68*) detailed electrochemical data for the processes

$$Fe(R_2dtc)_3^{-1} \rightleftharpoons Fe(R_2dtc)_3^0 \rightleftharpoons Fe(R_2dtc)_3^{+1}$$

with a great variety of substituents. As for Mn, a linear relation between the $E_{1/2}$ values for both electron transfers is observed. But surprisingly no influence of the spin state on the redox potentials is detected.

Apart from the six-coordinated Fe(III) compounds, mentioned above, also stable five-coordinate complexes $XFe(R_2dtc)_2$ (X = Cl, Br) are reported (*64*). Crystal structure analysis shows $ClFe(Et_2dtc)_2$ to have a square pyramidal geometry (*65*) as is found for $(NO)Fe(Me_2dtc)_2$ (*66*). Comparing the Fe–S bond lengths of both complexes, it is surprising to see that these distances do not differ significantly (2.32 Å and 2.30 Å, respectively), whereas the N–O bond distance of 1.02 Å is usually regarded as the value for a NO^+ group rather than NO or NO^-, pointing to a formal oxidation state of + 1 for the metal in $(NO)Fe(Me_2dtc)_2$. The metal-sulfur bond lengths suggest that these bonds are unaffected by the change in the formal oxidation state. This is also found for $Ni(R_2dtc)_2$ and $[Ni(R_2dtc)_3^+]$, but this could be caused by the difference in coordination symmetry. In contrast, the metal-sulfur distances in $Cu(R_2dtc)_2$ are longer than those in $[Cu(R_2dtc)_2^+]$.

The spin state of the compounds $XFe(R_2dtc)_2$ is 3/2 (*64*). Mössbauer spectra of $ClFe(Et_2dtc)_2$ in solution are almost identical with the spectrum of the six-coordinated $Fe(Et_2dtc)_3$ in solution. The similarity in magnetic susceptibility and in the isomer shift and quadrupole splitting parameters suggests a geometrical correspondency in solution, which can be attained by the binding of a solvent molecule to the sixth coordination site of the $ClFe(Et_2dtc)_2$ monomer (*67*).

Very recently, *Chant et al.* (*68*) reported the existence of a bis-chelated Fe(III) cation, obtained by titrating $Fe(i\text{–}Pr_2dtc)_3$ with Fe(III)-ions to the endpoint corresponding with $[Fe(i\text{–}Pr_2dtc)_2^+]$. The same cation can be obtained by the perchlorate acid hydrolysis of $Fe(i\text{–}Pr_2dtc)_3$.

Fe(IV) complexes containing $[Fe(R_2dtc)_3^+]$ ions were reported a few years ago by *Pasek* and *Straub* (*51*). Air oxidation of $Fe(R_2dtc)_3$, in the presence of gaseous BF_3 results in the formation of $Fe(R_2dtc)_3BF_4$. More recently, anaerobic oxidation with $HClO_4$ was reported (*68*).

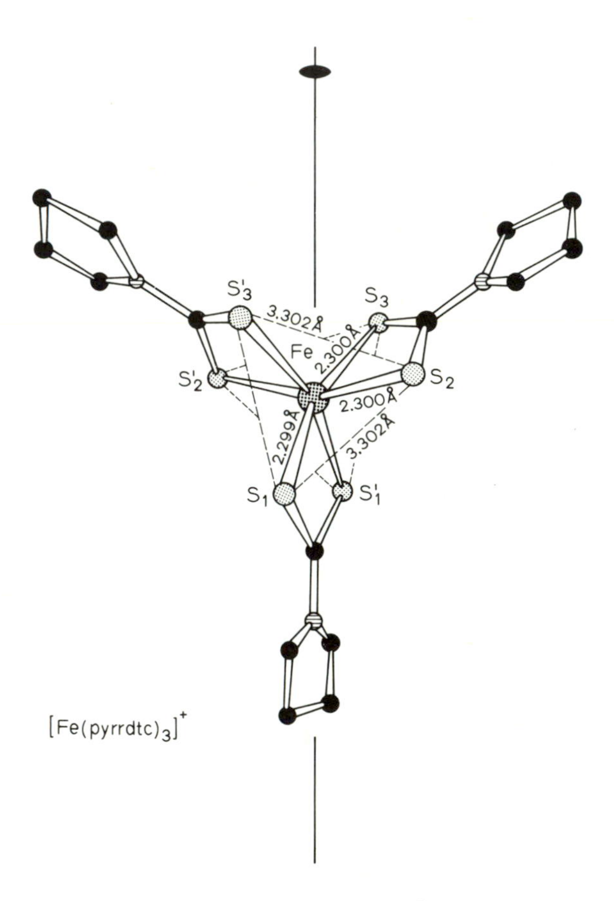

Fig. 4. The molecular structure of the $Fe[(CH_2)_4dtc]_3^+$ ion. (from ref. 69).

The compounds are 1 : 1 electrolytes in nitromethane and have magnetic moments of 3.2–3.4 BM at room temperature, which is slightly lower than is expected for a spin-paired d^4 electron configuration in an octahedral environment. Mössbauer and infrared data (especially the C–N stretching frequency) point to a higher charge on iron than in the Fe(III) compounds. *Golding et al.* (*52*) prepared Fe[$(CH_2)_4$*dtc*]$_3ClO_4$, of which the crystal structure was published recently (see Fig. 4) (*69*). The structure of the cation is similar to that of Fe[$(CH_2)_4$*dtc*]$_3$ (*63*). The FeS_6 core has neither the ideal octahedral geometry nor a trigonal prismatic arrangement. The twist angle is 38°. The average Fe–S distances in the low-spin Fe(IV) complex (2.30 Å) are significantly shorter (0.11 Å) than those in the high-spin Fe(III) dibutyl (*62*) and tetramethylene (*63*) complexes. This is ascribed (*69*) to the depopulation of antibonding e orbitals, *viz.* Fe(III) $(t_2^3\, e^2)$ → Fe(IV) (t_2^4) and the increasing oxidation state of the central atom. The latter contribution to the shortening appears to be minimal, since similar Fe–S bond lengths are observed for low-spin Fe(III) (t_2^5) (*63, 70*) and the present Fe(IV) (t_2^4) dithiocarbamates.

Ruthenium, Osmium

Ruthenium dithiocarbamates are synthesized with the metal in the oxidation state + 3.

The common type of *Ru(III)* containing complexes is Ru(R_2*dtc*)$_3$ (*71, 72*). The diethyl compound is isomorphous with Fe(Et_2*dtc*)$_3$ (*73, 74*). The oxidation potential of this complex in dmf (*75*) and acetonitrile (*76*) leads to the suggestion that the oxidation of this complex can easily be performed, and air oxidation in the presence of BF_3 is reported to yield Ru(Et_2*dtc*)$_3BF_4$ (*76*). However, a recent crystal structure analysis of a complex prepared in this way and recrystallised from an acetone-ether mixture revealed that no Ru(IV) species was present (*77*). Instead a complex Ru_2(Et_2*dtc*)$_5BF_4 \cdot C_3H_6O$ was found, in which two Ru(III) atoms are bridged by two sulfur atoms of the dithiocarbamato ligand. The existence of a Ru(IV) species thus seems questionable.

The osmium dithiocarbamato chemistry is virtually unexplored.

Cobalt

Of the oxidation states + 2, + 3 and + 4, found for the metal in cobalt dithiocarbamato complexes, + 3 is the most stable.

Co(II) complexes, which exist as Co(R_2*dtc*)$_2$ only, are in general unstable and oxidise readily to Co(III) compounds (*42*). The stability improves with larger alkyl groups. Co(*i*–Am_2*dtc*)$_2$ (*78*), Co[$(CH_2)_5$*dtc*]$_2$ and Co[S$(C_2H_4)_2$*dtc*]$_2$ (*79*) are reported to be rather stable, but their characterisation is far from complete.

The common type of *Co (III)* complexes is the easily obtainable $Co(R_2dtc)_3$ (*2*), which is diamagnetic and has a D_3 symmetry (*80*). The twist angle φ in $Co(Et_2dtc)_3$ is 43°.

In voltametric studies of $Co(R_2dtc)_3$ in various solvents, a one-electron oxidation and reduction step was detected (*75, 76, 162*).

A remarkable complex with the composition $Co_2(R_2dtc)_5BF_4$ resulted from an attempt of *Hendrickson* and *Martin* (*81*) to oxidise $Co(R_2dtc)_3$. These authors report diamagnetic, ionic complexes, in which the metal atoms are bridged by two sulfur atoms of the ligands, as is found for $Ru_2(Et_2dtc)_5BF_4$ (*77*).

In contrast, a *Co (IV)* complex $Co(Et_2dtc)_3BF_4$ was reported by *Gahan* and *O'Connor* (*76*) as well as by *Saleh* and *Straub* (*53*) as a result of the oxidation of $Co(Et_2dtc)_3$. The latter authors prepared, apart from the diethyl compound, the dicyclohexyl complex.

$Co(c{-}Hx_2dtc)_3BF_4$ is a 1 : 1 electrolyte in nitromethane and has a magnetic moment of 3.48 BM at room temperature, which is in between the value for a high-spin and a low-spin complex.

The authors suggest a mixed-spin state, analogous to that of the isoelectronic $Fe(R_2dtc)_3$.

Rhodium, Iridium

For rhodium the oxidation state +3 for the metal is normal, although complexes with the metal in the +4 and even the +5 state are reported.

The most common type of the *Rh (III)* complexes is the diamagnetic $Rh(R_2dtc)_3$ (*82*), which has presumably the same structure as $Co(R_2dtc)_3$(*72*), but also the dimeric $[Rh(Et_2dtc)_2Cl]_2$ is known (*83*).

Connelly et al. (*84*) reported the preparation and properties of a complex $[Rh(NO)(PPh_3)_2(Et_2dtc)]PF_6$. This compound, however, recently turned out to be $[Rh(PPh_3)_2(Et_2dtc)_2]PF_6$ (*85*), analogous to the earlier reported $[Rh(PPh_3)_2(Et_2dtc)_2]BF_4$ (*86*). The 1519 cm^{-1} absorption accordingly has to be ascribed to the C–N frequency.

Attempts to obtain *Rh (IV)* complexes by oxidation of $Rh(Et_2dtc)_3$ are reported to be successful (*76*), although the oxidation potential of the starting compound is as high as 1.06 V in acetonitrile. The characterisation of the obtained $Rh(Et_2dtc)_3BF_4$ is, however, far from complete.

The preparation of a seven-coordinated, formal *Rh (V)* complex $[Rh(NO)(Me_2dtc)_3]PF_6$ is claimed (*84*). As an infrared absorption of 1545 cm^{-1}, assigned to the N–O stretching vibration, is too low for a NO^+ group, the authors describe the NO group as uni-negative. This description leads to a formal oxidation state of +5 for the metal. Further investigations of this compound are in progress (*85*).

The chemistry of iridium dithiocarbamates is practically unexplored, and apart from $Ir(R_2dtc)_3$, which is presumably isostructural with $Co(R_2dtc)_3$ (*82, 87*), only the dimeric $[Ir(Et_2dtc)_2Cl]_2$ is reported (*83*).

Nickel

In nickel dithiocarbamato complexes the metal can have the formal oxidation state +2, +3 and +4.

Neutral, diamagnetic *Ni(II)* complexes with the formula $Ni(R_2dtc)_2$ are found (*2*). Structural studies for a great variety of R groups (*88–93*) showed a square planar coordination geometry as is expected for a four-coordinated metal in a d^8 configuration.

The oxidation and reduction of $Ni(R_2dtc)_2$ have been studied by voltametric techniques, the observed processes were irreversible (*162, 163*). $[Ni(Et_2dtc)_3^-]$ is detected in electrochemical studies as the reduction product of $[Ni(Et_2dtc)_3^+]$. It is, however, very unstable (*94*).

Ni(III) compounds, obtained by the oxidation of $Ni(R_2dtc)_2$ with mild oxidising agents are scarce. Only two examples are known, $Ni(Et_2dtc)_3$ and $Ni(Bu_2dtc)_2I$.

An EPR spectrum of a mixture of $Ni(Et_2dtc)_2$ with an 80-fold excess of Et_4tds showed a resonance signal, ascribed to $Ni(Et_2dtc)_3$ (*95*). The signal disappears upon heating the solution, which could be due to the formation of a Ni(IV) species. Though there is no proof to substanciate this, the alternative, the formation of a Ni(II) complex, seems less likely in view of the large excess of oxidant present. Electrochemical studies have revealed that $Ni(R_2dtc)_3$ is formed by cathodic reduction of $[Ni(R_2dtc)_3^+]$. It can be oxidised, forming the Ni(IV) compound again. The stability of $Ni(R_2dtc)_3$ is, however, low. It is decomposed in about one minute, probably disproportionating to Ni(II) and Ni(IV) complexes (*94, 96*).

Treating $Ni(Bu_2dtc)_2$ with iodine at $-30\,^\circ C$ yields an isolable product of composition $Ni(Bu_2dtc)_2I$ (*97*). The C–N stretching frequency of this compound is $1518\ cm^{-1}$, a value between that of $Ni(Bu_2dtc)_2$ ($1502\ cm^{-1}$) and $Ni(Bu_2dtc)_3I_3$ ($1529\ cm^{-1}$) (*98, 99*). Magnetic measurements showed that the Curie-Weiss Law is followed in the range 103–295 K, with a Weiss temperature $\Theta = -28$ K and a μ_{eff} of 1.33 BM. This value, low for a metal in a d^7 configuration, presumably results from a partly disproportionation into Ni(II) and Ni(IV) compounds.

A *Ni(IV)* complex, containing the $[Ni(Bu_2dtc)_3^+]$ ion is obtained when $Ni(Bu_2dtc)_2$ is treated with iodine at room temperature (*99*). Compounds with the same cation can easily be prepared by the direct reaction of $Ni(Bu_2dtc)_2$ with other halogens (*100, 101*) or $FeCl_3$ (*99, 52*). In these complexes the C–N stretching frequencies are increased with respect to the divalent complexes by approximately $30\ cm^{-1}$. The same effect is found for the analogous nickel diselenocarbamato complexes (*102*). Crystal structure elucidation of $Ni(Bu_2dsc)_3Br$ (*103*) and shortly thereafter, of the isomorphous $Ni(Bu_2dtc)_3Br$ (*104, 101*), showed the assigment of a formal oxidation number +4 for the metal to be correct, as no ligand oxidation to thiuram disulfide was found. No significant difference has been observed in the metal-sulfur distance, neither going from $Ni(Bu_2dsc)_2$ (*105*) to $[Ni(Bu_2dsc)_3^+]$, nor from $Ni(Et_2dtc)_2$ (*99*) to $[Ni(Bu_2dtc)_3^+]$. In contrast to a shortening in the metal-sulfur bond length, found upon oxidation of $Cu(R_2dtc)_2$ to $[Cu(R_2dtc)_2^+]$, such a decrease, which would be expected upon oxidation of $Ni(R_2dtc)_2$, is probably compensated by the change in coordination geometry. The structure of $Ni(Bu_2dtc)_3Br$ and $Ni(Bu_2dsc)_3Br$ consists of columns of closely packed S or Se atoms, respectively. The nickel atoms occupy half of the octahedral interstices in each column, and apart from the fact that no Ni–Ni interaction is present, a similarity with the nickel arsenide structure can be recognised. There is a close resemblance between the electronic spectra of $[Ni(Bu_2dtc)_3^+]$, $[Ni(Bu_2dsc)_3^+]$ and the isoelectronic compounds $Co(Bu_2dtc)_3$ and $Co(Bu_2dsc)_3$ (*102*) as could be expected. The Δ values for the nickel complexes are higher than those for the cobalt complexes. The values for B and therefore for β_{35} are extremely low in the nickel compounds. This indicates a considerable covalent character in the metal-ligand bond, mixing the *d*-orbitals with appropriate ligand orbitals.

Nigo et al. claim to have prepared $Ni(Et_2dtc)_2Br_2$ (*106*), whereas Jensen reports the preparation of $Ni(Et_2dsc)_2Br_2$ (*107*). Considering the IR, electronic and NMR spectral data, we believe these products to be $Ni(Et_2dtc)_3Br$ and $Ni(Et_2dsc)_3Br$ (*100*), respectively, contaminated with $NiBr_2$.

Apart from the reactions of $Ni(R_2dtc)_2$ or $Ni(R_2dsc)_2$ with halogens, remarkable features of these compounds are observed.

Upon standing in a $CHCl_3$ solution $Ni(Bu_2dsc)_2$ is converted into $Ni(Bu_2dsc)_3Cl$ (*102*) and the reaction of $Ni(Bu_2dtc)_2$ with an excess of $ZnCl_2$ or $ZnBr_2$ in dry, peroxide-free diethylether with exclusion of air ultimately yields $Ni(Bu_2dtc)_3X$ (X = Cl, Br) (*108*). The nature of these reactions is unknown at the moment.

The reaction of $Ni(PhHdtc)_2$ with Na_2S_2 does not lead to a Ni(IV) product as postulated by *Hieber* and *Brück* (*109*). A sulfur-rich dithiocarbimate of formula

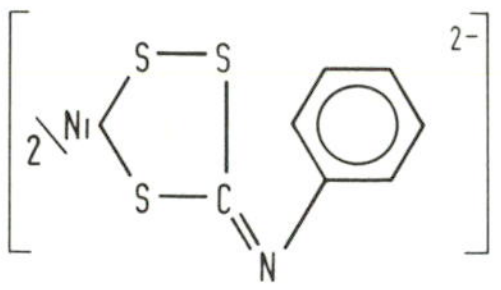

is obtained instead (*110*).

Palladium, Platinum

For palladium and platinum, dithiocarbamato complexes with the metal in the oxidation state of + 2 and + 4 are known.

The *Pd(II)* and *Pt(II)* complexes $M(R_2dtc)_2$ (*111, 46*) are diamagnetic and have a square planar geometry (*112, 113*) like the nickel analogues. In CH_2Cl_2 they can be oxidised irreversibly on a rotating platinum electrode, the Pd compound at a higher potential than the Pt compound (1.21 and 0.92 V vs s.c.e., respectively) (*150*).

A complex $[MX(R_2dtc)]_2$ is known for M = Pd (*114, 102*) but not for M = Pt. On the basis of infrared and Raman data (*102*), the structure is believed to have bridging halogen atoms between the metal atoms, the coordination geometry around each palladium being planar and the PdX_2Pd units folded about the X ... X line as is found for $[Rh(CO)_2Cl]_2$ (*115*).

Two types of complexes with *Pd(IV)* and *Pt(IV)* are known. The first type consists of neutral molecules $M(R_2dtc)_2X_2$ (*116*). They can be obtained by the direct reaction of the dithiocarbamato complexes $M(R_2dtc)_2$ with one mole of halogen. This is in contrast to the halogenation of $Ni(R_2dtc)_2$ at room temperature, for which $Ni(R_2dtc)_3X$ is obtained, perhaps formed from a kinetic labile intermediate $Ni(R_2dtc)_2X_2$. The slow substitution rate of octahedral Pd(IV) and Pt(IV) complexes accounts for the existence of the compounds $M(R_2dtc)_2X_2$. From these compounds the most well-investigated species is $Pt(Bu_2dtc)_2I_2$. Two products of this stoichiometry are found, a cis and a trans isomer. The trans product is labile in solution and is slowly converted into the cis product. The ESCA spectra of both isomers are the same and show only one absorption band present at the $S(2p_{1/2}, 2p_{3/2})$ binding energy of 165 eV, somewhat broadened due to unresolved $2p_{1/2}$ and $2p_{3/2}$ absorption. In $Pt(Bu_2dtc)_2$ one band is found at 164 eV. In the region of the $S(2s)$ ejection energy the same pattern is observed. Comparing these spectra with those of thiuram disulfide complexes, e.g. $HgCl_2(Bu_4tds)$, where two resolved peaks are present at 164 and 165 eV, due to the S atoms in two different environments, the conclusion can be drawn that in the $Pt(Bu_2dtc)_2I_2$ compounds the ligand is present as dithiocarbamate and not as thiuram disulfide. This is confirmed by the ESCA spectra recorded in the region of the $Pt(4f_{5/2}, 4f_{7/2})$ binding energy which show absorptions at 79.3 and 76.0 eV, compared with 77.0 and 73.6 eV for $Pt(Bu_2dtc)_2$, which points to the oxidation of the metal. The results of the X-ray structure analysis of cis-$Pt(Bu_2dtc)_2I_2$ show the platinum atom to be in a distorted octahedral coordination with two iodine and four

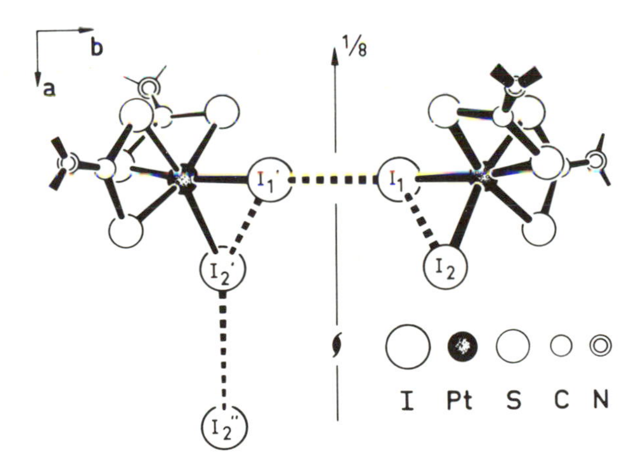

Fig. 5. Projection of two molecules of cis-Pt $(Bu_2dtc)_2I_2$ along the *c*-axis (omitting the butyl chains). The positions of a twofold axis, a twofold screw axis and one iodine atom of a third molecule are shown.

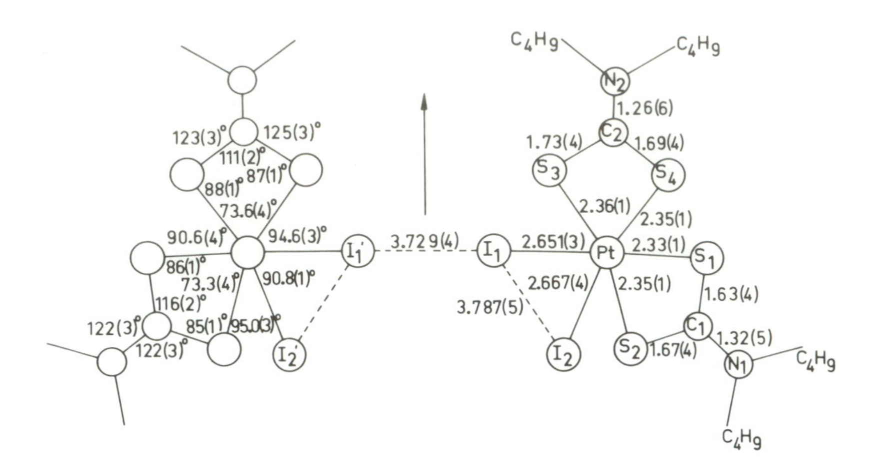

Fig. 6. Bond angles (degrees) and distances (Å) with e.s.d.'s of the cis-Pt $(Bu_2dtc)_2I_2$.

sulfur atoms (*116*) (see Fig. 5 and 6). The intramolecular iodine distance (3.787(5) Å) is intermediate between the length of an I–I bond (2.68 Å) in I_2 and an I–I van der Waals contact (4.30 Å) suggesting a three-centre bond PtI_2. One of the intermolecular I–I distances is also very short (3.729(4) Å) compared with the normal van der Waals contact, indicating a weak intermolecular binding. Pairs of molecules have I–I distances of 4.394(5) Å. In contrast to the dimeric character in the solid state, the complex is monomeric in solution.

The second type of *Pd(IV)* and *Pt(IV)* complexes have the general stoichiometry $M(R_2dtc)_3X$ (*117*). These compounds are isostructural with $Ni(Bu_2dtc)_3Br$ (*118*). However, in contrast to the preparation of the latter complex, the former complexes cannot be obtained by the direct halogenation of the divalent metal dithiocarbamates, so other preparative routes, as given in Table 1 must be followed.

In accordance with the general experience, the oxidation of Pd(II) is more difficult than that of Ni(II) and Pt(II). Oxidation of Ni(II) and Pt(II) is possible with $Br_2Cu(R_2dtc)$ and even with $CuBr_2$, e.g. $Pt(Bu_2dtc)_2 + 2\,CuBr_2 \rightarrow Pt(Bu_2dtc)_2Br_2 + 2\,CuBr$ (*102*), such reactions do not occur for $Pd(R_2dtc)_2$ (Table 1). The Pd(IV) compounds are less stable than the analogous Ni(IV) and Pt(IV) compounds.

The ionic $[Pt(Bu_2dtc)_3^+]$ complexes in $CHCl_3$ solution have electronic spectra quite similar to those of $[Ni(Bu_2dtc)_3^+]$ complexes. As is expected, the C–N stretching frequency of $[M(Bu_2dtc)_3^+]$ (M = Ni, Pd, Pt) is increased with respect to the complexes $M(Bu_2dtc)_2$ (*116, 117*). Instead of diamagnetism, a rather high susceptibility is found for $M(Bu_2dtc)_3X$ complexes (for instance $\chi_{mol}^{para} = 1191 \cdot 10^{-6}$ cgsu for $Pt(Bu_2dtc)_3I$). This susceptibility is nearly temperature-independent. No explanation for this phenomenon can be given at the moment (*117*).

Table 1

Reactants		Resulting complexes Ni	Pd	Pt
$M(Bu_2\mathit{dtc})_2X_2$	$Bu_4\mathit{tds}$		$Pd(Bu_2\mathit{dtc}_3)X$	$Pt(Bu_2\mathit{dtc})_3X$
$M(Bu_2\mathit{dtc})_2$	$Br_2Cu(Bu_2\mathit{dtc})$	$Ni(Bu_2\mathit{dtc})_3Br$	no reaction	$Pt(Bu_2\mathit{dtc})_3Br$
$M(Bu_2\mathit{dtc})_2$	$(Bu_4\mathit{bitt})Cu_2X_6$ [a]	$Ni(Bu_2\mathit{dtc})_3X$ [b]	$[Pd(Bu_2\mathit{dtc})_3]_2Cu_2X_6$ [c]	$Pt(Bu_2\mathit{dtc})_3CuX_2$

[a]) Excessive halogenation of several dithiocarbamato and thiuram disulfide complexes yields compounds in which the ligand is oxidised to the dipositive 3,5-*B*is (N,N dialkyl*i*minium) *t*ri*t*hiolane-1,2,4 cation $R_2N^+ = C{-}S{-}S{-}C = N^+R_2$ (the two C atoms bridged by S), abbreviated $[R_4\mathit{bitt}^{2+}]$. Depending on the halogen used, one atom sulfur per two dithiocarbamato units is removed as S_2X_2 (X = Cl, Br) or as elemental sulfur (X = I) (*99, 102, 117, 119, 120*).

[b]) So far no $[Ni(Bu_2\mathit{dtc})_3^+]$ complexes are found with any halometallate anion apart from $[FeX_4^-]$ (X = Cl, Br).

[c]) Decomposes in $CHCl_3$ solution into $Pd(Bu_2\mathit{dtc})_2$ and $X_2Cu(Bu_2\mathit{dtc})$.

Group I Transition Metal Dithiocarbamato Complexes

Copper

In copper dithiocarbamato complexes the metal can have the oxidation states +1, +2 and +3. The complexes synthesized up to the present are summarised in Table 2.

The reaction of metallic copper with thiuram disulfides yields complexes of *Cu(I)*, which are polymeric in solution as well as in the solid state (*121, 122*). In $[Cu(Et_2dtc)]_4$ (*123*) the copper atoms are located at the corners of a slightly distorted tetrahedron with Cu–Cu distances ranging from 2.6–2.7 Å. Each of the copper atoms is coordinated to three sulfur atoms in a nearly planar triangular arrangement and each sulfur atom coordinates one or two copper atoms.

Refluxing $[Cu(Et_2dtc)]_4$ in $CHCl_3$ yields a compound which analyses as $Cu_3(Et_2dtc)_2Cl$ (*124*). The analogous butyl compounds could be prepared by the reaction of 1 mole $[Cu(Bu_2dtc)]_4$, 2 moles Cu and 1 mole X_2 (X = Cl, Br) in CS_2. Molecular weight determinations in benzene show an association factor 3 and infrared data point to terminal halogen atoms (*99*). No crystal structure data of these compounds are available.

For *Cu(II)*, stable, neutral complexes $Cu(R_2dtc)_2$ are easily prepared (*46, 125*). The properties of these complexes are well studied (*126, 132*). Although monomeric in solution (*126, 133*), they are dimeric in the solid state and a structural study of $Cu(Et_2dtc)_2$ (*134*) shows that the copper atom lies 0.26 Å out of the plane formed by four sulfur atoms at a distance of 2.30(1) Å. A fifth, long Cu–S bond (2.85 Å) is approximately perpendicular to this plane, whereas a hydrogen atom of an ethyl group is situated at the other side of the S_4 plane at a distance of 2.86 Å from the copper atom.

From the reaction of equimolar amounts of CuX_2 and $Cu(Et_2dtc)_2$, $XCu(Et_2dtc)$ is formed (*135, 99*), which can be obtained also by the oxidation of $[Cu(Et_2dtc)]_4$ with X_2 in stoichiometric quantities (*99*). A remarkable feature of these complexes is their

Table 2. Types of copper dithiocarbamato complexes found so far [a]):

Oxidation number	Compound		
I	$[Cu(R_2dtc)]_4$	$Cu_3(R_2dtc)_2X$	
II	$[Cu(R_2dtc)_2]_2$	$[XCu(R_2dtc)]_2$	
III	$[Cu(R_2dtc)_2^+]X^-$	$X_2Cu(R_2dtc)$	
II, I	$Cu(R_2dtc)_2 \cdot nCuBr$	$Cl_3Cu_3(R_2dtc)_2$	
III, I	$[Cu(R_2dtc)_2^+]\ [Cu_{n-1}Br_n^-]$	$[Cu(R_2dtc)_2^+]_2[Cu_2Br_4^{2-}]$	
III, II	$[Cu_3(R_2dtc)_6^+]\ [MX_3^-]_2$	$Cu_2(R_2dtc)_4I_3$	$Cu_2(R_2dtc)_3Br_2$

[a]) See text for R, X and n.

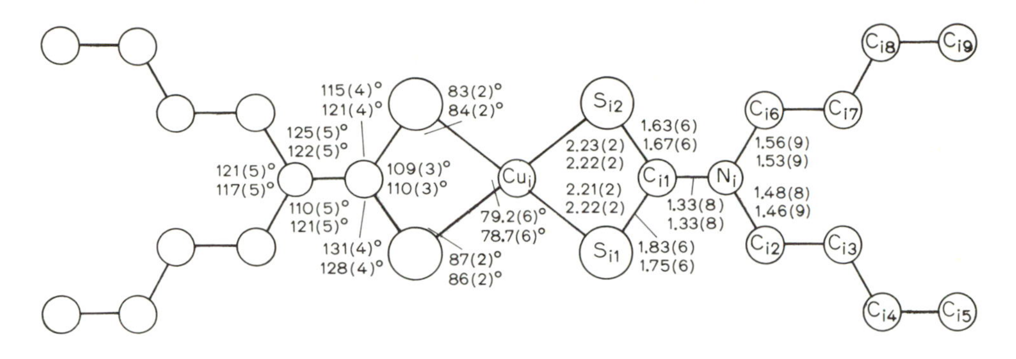

Fig. 7. Bond angles (degrees) and distances (Å) of the $[Cu(Bu_2dtc)_2^+]$ ion with e.s.d.'s.

conductivity in nitrobenzene solution. A good explanation of this phenomenon could not be given until now. From the infrared data the compounds were suggested to be dimeric, which was confirmed recently by a crystal structure elucidation of the chlorine compound (*136*). This study reveals a square planar coordination geometry around the copper atoms, which are bridged by chlorine atoms.

Two classes of complexes with *Cu(III)* are known. The first type contains the cation $[Cu(R_2dtc)_2^+]$, the second type has the general formula $X_2Cu(Bu_2dtc)$ (X = Cl, Br).

The reaction of I_2, $FeCl_3$ or $Fe(ClO_4)_3 \cdot 6\,H_2O$ with $Cu(R_2dtc)_2$ yields $[Cu(R_2dtc)_2^+]X^-$ ($X^- = I_3^-$, $[FeCl_4^-]$ or $[ClO_4^-]$, respectively) (*98, 52*).

It seems obvious that complexes with $[Cu(R_2dtc)_2^+]$ can be isolated only when the anion is large. This is supported by the reactions of $Cu(Bu_2dtc)_2$ or $[Cu(Bu_2dtc)]_4$ with the adequate amounts of Cl_2 or Br_2 for which, instead of $Cu(Bu_2dtc)_2X$, $X_2Cu(Bu_2dtc)$ is formed (*137*). *Nigo et al.* described the preparation of $I_2Cu(Bu_2dtc)$ (*106*). In spite of many attempts, however, we could not reproduce this compound.

Both types of diamagnetic d^8 Cu(III) complexes are square planar. They show a remarkable shortening of the Cu–S distances with respect to those in $Cu(Et_2dtc)_2$ (2.30(1) Å in $Cu(Et_2dtc)_2$ (*134*), 2.22(2) Å in $Cu(Bu_2dtc)_2I_3$ (*138*) and 2.19(1) Å in $Br_2Cu(Bu_2dtc)$ (*137*)). Moreover, the Cu–Br bond length of 2.31(1) Å in $Br_2Cu(Bu_2dtc)$ is short compared with those in Cu(II)–Br compounds (2.40–2.46 Å). As all interatomic distances in the ligands remain unaffected, the oxidation occurs locally in the CuS_4 and Br_2CuS_2 moieties, respectively. This is in accord with MO calculations (vide infra), which showed that the unpaired electron in $Cu(R_2dtc)_2$ is in a molecular orbital of predominant metal character.

Voltametric studies (*150, 162, 163*) have revealed that the electron transfers $Cu(R_2dtc)_2^{-1} \rightleftharpoons Cu(R_2dtc)_2^0 \rightleftharpoons Cu(R_2dtc)_2^{+1}$ are reversible at a rotating platinum electrode.

An interesting phenomenon is the occurrence of dithiocarbamates with copper in mixed oxidation states.

The combination of *Cu(III)* and *Cu(II)* is found in the complexes $Cu_3(Bu_2dtc)_6(MBr_3)_2$ (M = Zn, Cd, Hg) with copper in the average oxidation state + 8/3, and $Cu_2(Bu_2dtc)_4I_3$ and $Cu_2(R_2dtc)_3Br_2$ (R = Me, Et), both with copper in the average oxidation state + 5/2.

The complexes $Cu_3(Bu_2dtc)_6(MBr_3)_2$ (*139*) dissociate in solution into two ions $[Cu(Bu_2dtc)_2^+]$, one molecule $Cu(Bu_2dtc)_2$ and two ions $[MBr_3^-]$ for M is Zn and Hg, or one ion $[M_2Br_6^{2-}]$ for M is Cd. In the solid state the compounds follow the Curie-Weiss Law in the temperature range 100–293 K and the molar magnetic susceptibility corresponds to one unpaired electron. From EPR single crystal spectra the principal g and A values suggest the electron to be localised on one copper atom only. The values correspond well with those found for a solid solution of monomeric $Cu(Et_2dtc)_2$ in $Ni(Et_2dtc)_2$, as given by *Weeks* and *Fackler* (*130*). Evidence for the existence of copper in two oxidation states in the solid state is given by a crystal structure determination (*139*).

The ion $[Cu_3(Bu_2dtc)_6^{2+}]$ is centrosymmetric, containing three $Cu(Bu_2dtc)_2$ units, a central Cu(II) $(Bu_2dtc)_2$ is sandwiched between two Cu(III) $(Bu_2dtc)_2$ layers. The former unit is planar, centrosymmetric with Cu–S distances of 2.30(1) and 2.35(1) Å, which are not significantly different from those in $Cu(Et_2dtc)_2$ (2.30(1) Å). The centrosymmetric copper atom has, apart from the four Cu–S bonds already mentioned two longer Cu–S bonds at 2.88 Å, thus reaching a pseudo-octahedral coordination, which is not unusual for a d^9 configuration. The other two equivalent $Cu(Bu_2dtc)_2$ units are slightly deformed planes with much shorter Cu–S distances (2.22(1) Å), which are the same as those in $Cu(Bu_2dtc)_2I_3$ (*22*). Expecting a planar structure for a low spin d^8 configuration it is surprising to see that the copper is lifted out of the plane of the four sulfur atoms by 0.26 Å, whereas a fifth, long Cu–S bond is present, which has a bond length of 2.88 Å, values both corresponding with those found for the copper atom in $Cu(Et_2dtc)_2$.

Some infrared frequencies are given in Table 3 together with those for $[Cu(Bu_2dtc)_2^+]$ $[MBr_3^-]$ and $Cu(Bu_2dtc)_2$. The two C–N and two Cu–S stretching frequencies clearly can be correlated with copper in the oxidation states + 2 and + 3, respectively. The frequencies related with the Cu(II) unit are increased with respect to $Cu(Bu_2dtc)_2$, those of the Cu(III) units show a slight decrease with respect to those of the corresponding $[Cu(Bu_2dtc)_2^+]$ complexes.

Table 3. Some infrared frequencies in cm^{-1} of $Cu_3(Bu_2dtc)_6(MBr_3)_2$.

	ν(C–N)		ν(Cu–S)	
$Cu_3(Bu_2dtc)_6(ZnBr_3)_2$	1512	1548	362	398
$Cu_3(Bu_2dtc)_6(CdBr_3)_2$	1511	1548	360	397
$Cu_3(Bu_2dtc)_6(HgBr_3)_2$	1510	1549	360	397
$Cu(Bu_2dtc)_2(ZnBr_3)$		1549		408
$Cu(Bu_2dtc)_2(CdBr_3)$		1551		411
$Cu(Bu_2dtc)_2(HgBr_3)$		1550		404
$Cu(Bu_2dtc)_2I_3$		1562		411
$Cu(Bu_2dtc)_2$	1500		352	

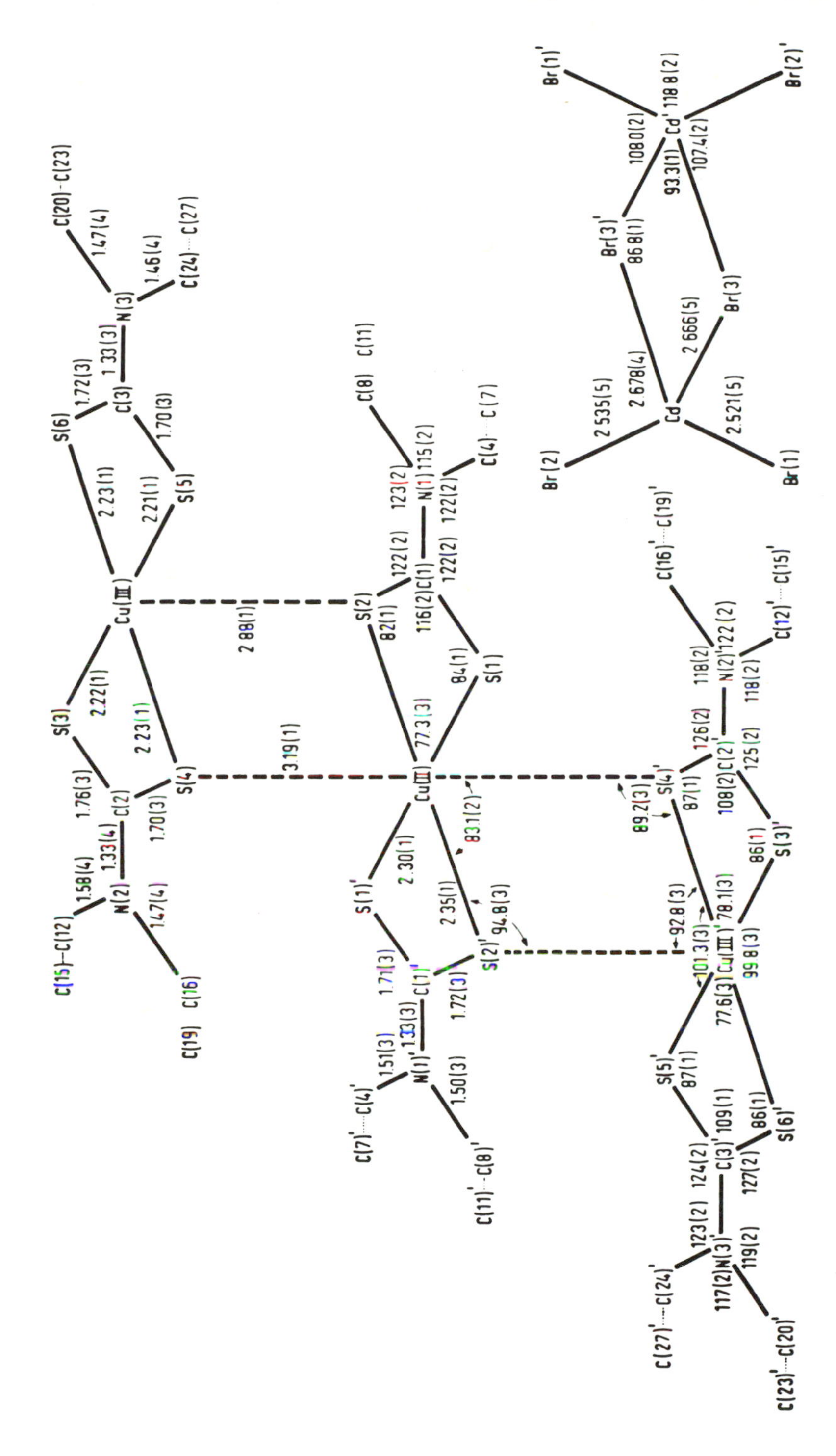

Fig. 8. Bond angles (degrees) and distances (Å) with e.s.d.'s of the important atoms in $[Cu_3(Bu_2dtc)_6]$ $[Cd_2Br_6]$.

$Cu_2(Bu_2dtc)_4I_3$(*99*) shows a magnetic moment at room temperature corresponding with one unpaired electron per two copper atoms, whereas the infrared spectra indicate the presence of $Cu(Bu_2dtc)_2$ and $[Cu(Bu_2dtc)_2^+]$ units. No further experimental data are available at present.

$Cu_2(R_2dtc)_3Br_2$ also has two C–N and two Cu–S stretching vibrations, corresponding with Cu(II) and Cu(III) species. Magnetic susceptibility measurements point to one unpaired electron per formula unit. ESCA spectra in the region of the $Cu(2p_{3/2})$ binding energy confirm the presence of two types of Cu atoms. X-ray structural data reveal the ethyl compound to consist of layers in which $Cu(Et_2dtc)_2$ and $Br_2Cu(Et_2dtc)$ units alternate. The Cu–S distances in the $Cu(Et_2dtc)_2$ unit have an average value of 2.20 Å, those in the $Br_2Cu(Et_2dtc)$ unit 2.28 Å, whereas the Cu–Br distance is 2.40 Å. Comparison of these values with those of Cu(III) complexes, shows that a formal oxidation number + 2 could be assigned to the copper atom in the $[Br_2Cu(Et_2dtc)^-]$ unit and a formal oxidation number + 3 to the copper atom in the $[Cu(Et_2dtc)_2^+]$ unit. The average distances between the units are Cu(II)–S 2.97 Å, Cu(III)–S 3.27 Å and Cu(III)–Br 2.97 Å. From these distances the conclusion can be drawn that both copper atoms are in a distored octahedral environment. The molar weight measured in $CHCl_3$ solution corresponds with the value for monomeric $Cu_2(Et_2dtc)_3Br_2$ (*140*).

Complexes of the type $[Cu(R_2dtc)_2^+]$ $[Cu_{n-1}Br_n^-]$ with *Cu(III)* and *Cu(I)* were recently prepared (*141*). The value of *n* is dependent on the length of the alkyl chain, *n* is 7 for butyl, 5 for propyl and ethyl and 3 for methyl. Infrared data point to the presence of the $[Cu(R_2dtc)_2^+]$ cation. All compounds are diamagnetic. ESCA data of $[Cu(Bu_2dtc)_2]$ $[Cu_6Br_7]$ reveal the presence of copper in two different oxidation states (933.8 and 932.5 eV for the $Cu(2p_{3/2})$ binding energy) in a ratio of 1 : 6 (*194*).

Furthermore, a compound $[Cu(Pr_2dtc)_2^+]_2[Cu_2Br_4^{2-}]$ could be prepared. It is remarkable that this composition so far exists only for R is propyl (*142*).

Finally the combination *Cu(II)* and *Cu(I)* is reported. This mixed oxidation state was found by *Golding et al.* (*143*) in the complexes $Cu[(CH_2)_5dtc]_2 \cdot 4\,CuBr$ and $Cu[(CH_2)_5dtc]_2 \cdot 6\,CuBr$ and in our studies of $Cu(Bu_2dtc)_2 \cdot 7/2\,CuBr$, $Cu(Pr_2dtc)_2 \cdot 2\,CuBr$ and $Cu(Et_2dtc)_2 \cdot 2\,CuBr$ (*141*). X-ray structure determination performed by Golding revealed the compound $Cu[(CH_2)_5dtc]_2 \cdot 4\,CuBr$ to be composed of polymeric sheets of individual $Cu[(CH_2)_5dtc]_2$ molecules, linked to polymeric CuBr chains via Cu–S bonds.

Recently, Taylor mentioned in a personal communication (*136*) the structure of the Cu(II)–Cu(I) compound $Cu_3Cl_3(Et_2dtc)_2$. The compound is polymeric and consists of layers of $Cu_2Cl_2(Et_2dtc)_2$ coupled by $[Cu_2Cl_2(Et_2dtc)]_2$ units (*195*).

Silver

In the silver dithiocarbamato chemistry only univalent metal complexes could be isolated until now. EPR studies, however, revealed that the oxidation state + 2 can be achieved in solutions. The existence of the oxidation state + 3 in solution is also reported.

The *Ag(I)* complexes $Ag(R_2$*dtc*$)$ were investigated by *Åkerström* (*121*). They are polymeric in solution as well as in the solid state. The crystal structure of $[Ag(Pr_2$*dtc*$)]_6$ contains discrete hexameric molecules. The metal atoms form a somewhat distorted octahedron with six comparatively short and six longer edges. The short edges correspond to metal-metal distances, which are comparable or somewhat longer than those in the metallic phase of silver. The long edges form two centrosymmetrically related triangles in the silver octahedron. Outside each of the other six faces of the octahedron one dithiocarbamato ligand is situated, linked to the silver atoms of the face by silver-sulfur coordination.

One of the sulfur atoms is linked to one and the other to two silver atoms. The silver coordination is threefold but not planar, the metal atoms being situated "inside" the plane of the coordinating sulfur atoms (*144*).

Attempts to oxidise silver dithiocarbamato complexes with halogens to compounds with the metal in higher oxidation states obviously failed. By addition of iodine to a solution of $[Ag(Bu_2$*dtc*$)]_6$ in $CHCl_3$, an insoluble product is formed with the composition $I_7Ag_{11}(Bu_2$*dtc*$)_4$(*141*). With other alkyl groups similar complexes are obtained. Investigations about the nature of this type of compounds are in progress.

Ag(III) dithiocarbamate, observed as a red-coloured solution, obtained by the reaction of one mole Bu_4*tds* and 1/6 mole $[Ag(Bu_2$*dtc*$)]_6$ in benzene, was reported by *Bergendahl* and *Bergendahl* (*145*). This product is rather unstable as the red solution turns to the blue-coloured $Ag(Bu_2$*dtc*$)_2$ within ten minutes.

Gold

The common oxidation states for gold in the dithiocarbamato complexes are + 1 and + 3. Like for the silver compounds the oxidation state + 2 is observed in diluted liquid and solid solutions only.

The *Au(I)* complexes $Au(R_2$*dtc*$)$ are dimeric in solution (*121*). The crystal structure analysis of $Au(Pr_2$*dtc*$)$ reveals that these compounds also are dimeric in the solid state (*146*). The most interesting feature of the X-ray analysis is the Au–Au distance of 2.76 Å, which is shorter than the metal-metal distance in metallic gold (2.88 Å (*147*)). Each gold atom is coordinated to a sulfur atom of two different dithiocarbamato ligands, the linear AuS_2 coordination being perpendicular to a twofold axis of symmetry, which passes through the metal atoms. Another twofold symmetry axis passes through the C–N bonds of both of the dithiocarbamate ligands. The overall symmetry of the molecule is D_2. Raman studies of the metal-metal bonding (*148*) suggest a contribution of resonance structures as

```
R          S⊖   Au    S          R
 \⊕       /     |     .\        /
  N ═══ C       |   ⊕ : C ─── N
 /        \     |     .'/       \
R          S⊖   Au    S          R
```

EPR studies are performed at the *Au(II)* complexes $Au(R_2dtc)_2$, stabilised by dilution in $Ni(R_2dtc)_2$ or $Zn(R_2dtc)_2$ host lattices (e.g. *149*).

Au(III) dithiocarbamato complexes exist, analogous to the copper compounds, in two types, ionic $[Au(R_2dtc)_2^+]$ complexes and non-ionic species $Br_2Au(R_2dtc)$; in addition, $Au(Bu_2dtc)_3$ has been reported.

$[Au(R_2dtc)_2^+]$-ion containing compounds can be prepared with different anions (*150, 151, 152, 153*). Remarkable is the existence of $[Au(Bu_2dtc)_2^+]Br^-$, indicating that small-sized anions are apt for the isolation of a cation $[Au(Bu_2dtc)_2^+]$, whereas they are not for the isoelectronic and isostructural $[Cu(Bu_2dtc)_2^+]$.

$[Au(Bu_2dtc)_2^+]\,[AuBr_4^-]$ can be converted at 41 °C into the isomeric, non-ionic $Br_2Au(Bu_2dtc)$ (*151*), which is isomorphous with $Br_2Cu(Bu_2dtc)$ (*137*).

A crystal structure analysis of the diamagnetic $Au(Bu_2dtc)_3$ revealed the gold atom to be planar coordinated by four sulfur atoms of three dithiocarbamato ligands, of which one is bidental and two monodental (*154*).

The electron transfer $Au(R_2dtc)_2^{+1} \rightleftharpoons Au(R_2dtc)_2^0$ has been detected by voltametric measurements (*163*). The half-wave potentials of the quasi-reversible process depends on the substituent R according to the Taft relation, as was described for Mo, W and Mn (*37*). The value of ρ decreases in the series $Au > Mn > Mo \approx W$, which indicates that in this sequence the mixing of ligand orbitals into the redox orbital decreases. The dominant ligand character of the unpaired electron MO in $Au(R_2dtc)_2$ relative to those in copper and silver compounds is found from Extended Hückel MO calculations, as will be discussed later on.

The only example of a compound with mixed oxidation numbers *Au(III)–Au(I)* is $[Au(Bu_2dtc)_2^+]\,[AuBr_2^-]$ (*155*), which can be obtained by the addition of one mole Br_2 or two moles $Br_2Au(Bu_2dtc)$ to one mole $[Au(Bu_2dtc)]_2$.

The combination of *Au(III)* and *Cu(I)* or *Ag(I)* is realised in the complexes $[Au(Bu_2dtc)_2]\,[CuBr_2]$ (*152*) and $[Au(Bu_2dtc)_2]\,[AgBr_2]$ (*156, 157*) from which the first is isomorphous with $[Au(Bu_2dtc)_2]\,[AuBr_2]$, whereas the second is not.

An interesting combination is *Au(III)–Cu(III)–Cu(II)* in the compound $[AuCu_2(Bu_2dtc)]_6[Hg_2I_6]$, isomorphous with $[Cu_3(Bu_2dtc)_6]\,[Cd_2I_6]$, in which one copper atom is situated at a centre of symmetry and one copper and one gold atom are statistically distributed among two equivalent general positions. The copper at the centre is in the oxidation state +2, the other copper and the gold atom are in the oxidation state +3 (*158*).

Apart from the mentioned Au(III) complexes, a wide range of alkylgold(III) dithiocarbamato compounds of the type $R'_2Au(R_2dtc)$ were investigated (*159, 160*).

Group IIb Metal Dithiocarbamato Complexes

Zinc, Cadmium, Mercury

In dithiocarbamato complexes of these elements the metal has, without exception, the oxidation state +2, and the compounds are of the type $M(R_2dtc)_2$ only (*2*).

Oxidation of these complexes with halogen results in the oxidation of the ligand, yielding a thiuram disulfide complex $X_2M(R_4tds)$, leaving the oxidation state of the metal unchanged (*99, 100, 161*).

McCleverty and Morrison synthesized complexes with anions of the type $[Zn(R_2dtc)_3^-]$ and $[X_2Zn(R_2dtc)^-]$(*193*). These species are interesting, as they belong to the few examples of anionic dithiocarbamates that could be isolated.

EPR Studies

The literature dealing with EPR studies of transition metal dithiocarbamato complexes is extensive. Interesting results were obtained about the interaction of copper compounds with various solvents (*165, 166, 167, 168*) and about dimer formation of $Cu(R_2dtc)_2$ in frozen solutions, (*133, 169*) whereas extensive EPR studies about other transition metal dithiocarbamato complexes are reported as well (*170, 5, 171, 37*). As the measurements of the planar d^9 systems are most suitable for comparison with theoretical studies, we shall pay attention to the results of these investigations on Cu(II), Ag(II) and Au(II).

The silver (II) and gold (II) dithiocarbamato complexes are stable only in liquid and solid solutions. Because such a dilution of the paramagnetic complex with diamagnetic solvent or host lattice molecules prevents electron spin-electron spin interactions, also the copper(II) complexes were studied in the diamagnetic host lattices of $Ni(R_2dtc)_2$ or $Zn(R_2dtc)_2$. The pure copper compound consists of dimeric units. In liquid solution or diluted in $Ni(R_2dtc)_2$, the compound is monomeric. If $Zn(R_2dtc)_2$ is used as a host lattice a $Cu(R_2dtc)_2$ and a host lattice molecule interact, resulting in a structure similar to that of $[Cu(R_2dtc)_2]_2$. By increasing the concentration of the paramagnetic ion in the $Zn(R_2dtc)_2$ host lattice, electron spin-electron spin interactions in well-defined copper-copper (*172*) and silver-silver (*173, 174, 149*) dimers could be studied.

The analysis of the spectra of $Au(R_2dtc)_2$ is difficult because of the large influence of the quadrupole moment of the gold nucleus (*149, 175, 176, 177*), the determination of the spin hamiltonian parameters from the copper and silver spectra is straightforward. It turned out that the interpretation of these experimentally determined spin hamiltonian parameters in terms of LCAO coefficients and MO energies is impossible for these complexes. This is caused mainly by the low symmetry of these compounds (C_i at the most) and by the substantial contributions of the sulfur atoms to the g-tensor (*178, 179, 180, 181*). Furthermore, terms higher than second order cannot be omitted for the calculation of the g-tensor of silver and gold complexes because of the strong spin-orbit coupling in the metal atoms (*149*).

Because these circumstances hamper the direct "translation" of the experimentally determined EPR parameters into LCAO coefficients and MO energies, a different method has been used to extract information about the electronic structure of these complexes: the measured EPR parameters are compared with those calculated from the LCAO coefficients and MO energies which were obtained with an iterative Extended Hückel method (*178, 179, 181*). The advantage of this procedure is that all metal and ligand orbitals can be included and that the systems may have any symmetry. A disadvantage of the Extended Hückel method, however, is that the hamiltonian matrix contains two empirical constants which may be chosen arbitrarily. Because the results of this MO method depend strongly on these empirical constants, it is necessary to find a criterion for the correctness of their values. To that end the experimental g-values and hfs's of $Cu(Et_2dtc)_2$ diluted in $Ni(Et_2dtc)_2$ were compared with the values, calculated from the results of the Extended Hückel method for varying values of the two empirical constants. It appeared that it was possible to calculate these nine EPR parameters in agreement with the experiment, employing reasonable values for the two empirical constants. These values for the empirical constants have been used in all further calculations on related compounds, as for instance $ClFe(Et_2dtc)_2$ (see Mössbauer spectroscopy) (*132*) and diselenocarbamato complexes of copper (*179, 180, 181*).

From the good correspondence between the calculated and experimental *g*-values and hfs's, one may conclude that the MO's, calculated with these values for the empirical parameters, give a fair description for the ground state of $Cu(Et_2dtc)_2$. The results of the calculation show that the bonding in $Cu(Et_2dtc)_2$ is largely covalent, with overlap populations between copper and sulfur atoms of 0.22 electron unit.

On trying to assign an oxidation number to the copper atom one finds that the atomic charges, as calculated with the Mulliken definition (*183*), are remarkably small: +0.01 on the copper atom and −0.28 on the sulfur atoms. Hence, from these values one should assign the oxidation number 0 to the copper atom.

On the other hand, this MO description matches a crystal field description of a Cu(II) ion insofar as the five highest occupied MO's are concerned: four of these are doubly occupied and have a predominant copper 3*d* character (65% at least), whereas the fifth orbital contains the unpaired electron and still has 50% copper 3*d* character. The very small Mulliken charge on the copper atom, which does not fit in the description of a Cu(II) ion, is caused by the partial occupation of the 4*p* orbitals, which effect is neglected in the crystal field theory.

Application of the same method on $Ag(Et_2dtc)_2$ and $Au(Et_2dtc)_2$ was unsuccessful: the agreement between the calculated and the experimentally determined EPR parameters was poor, probably because of an underestimation of the covalent character of the metal-sulfur bonds and because of incorrect excitation energies (*149*). Therefore, some information about the electronic structure was substracted directly from the experimental results, making approximations about the molecular symmetry, the ligand contributions to the *g*-tensor and second order contributions to the hfs. The LCAO coefficients, which are calculated in this way, indicate that the convalency in the metal-sulfur bonds increases very strongly, going from copper to silver and gold: the metal 3*d* character of the MO of the unpaired electron decreases from 50% in $Cu(Et_2dtc)_2$ to 26% in $Ag(Et_2dtc)_2$ and 15% in $Au(Et_2dtc)_2$. From these results one should conclude that the unpaired electron in $Au(Et_2dtc)_2$ is situated mainly on the ligands.

Mössbauer Spectroscopy

In Table 4 typical values are given for the isomer shift (IS) and the quadrupole splitting (QS) of dithiocarbamato complexes with iron in various formal oxidation states.

From these data it is obvious that neither the IS nor the QS is useful for the assignment of an oxidation number, although the divergence of the IS for a fixed oxidation number and a fixed spin multiplicity is smaller than that of the QS.

Table 4. IS and QS at 95 K of various iron dithiocarbamates

Formal metal oxidation number	Spin multipl.	IS [a], [b]	QS [a]	Example	Ref.
+ 1	1	0.59–0.70	0.64–0.83	$(NO)Fe(Et_2dtc)_2$	(*184*)
+ 2	3 or 5	1.12–1.22	3.76–4.16	$[Fe(Et_2dtc)_2]_2$	(*185*)
	5	1.26–1.30	2.26–3.11	$(Et_4N)Fe(Et_2dtc)_3$	(*185*)
+ 3	2 and 6	0.70–0.76	0.5 –1.0	$Fe(Et_2dtc)_3$	(*186*)
	4	0.74	2.40–3.04	$ClFe(Et_2dtc)_2$	(*67, 187, 188*)
+ 4	3	0.53–0.57	2.29–2.54	$Fe(Me_2dtc)_3ClO_4$	(*52*)
	3 ?	0.50–0.53	2.21–2.43	$Fe(Et_2dtc)_3BF_4$	(*51*)

[a] in mm/sec; [b] relative to $Na_2[Fe(CN)_5NO] \cdot 2\,H_2O$.

It is impossible to translate the IS in a 4*s* electron density, because of the shielding of the inner-core *s*-electrons by the 3*d* electrons, which depends on the covalency.

To examine the effect of covalency on the QS, the values of the QS of some compounds have been calculated with the aid of the Extended Hückel MO method (*182, 61*). In these calculations the values for the empirical parameters used were obtained from comparison with EPR experiments (see EPR studies). This method is suitable only for molecules with low symmetry, because effectsoof spin-orbit coupling and thermal mixing have been neglected.

Bis(diethyldithiocarbamato)iron(III)chloride

The method mentioned has been applied to $ClFe(Et_2dtc)_2$ first, because for this compound, and some related ones, an abnormally large QS has been observed, while the crystal field theory predicts QS = 0 for this intermediate spin (S = 3/2) compound. The calculated relative energies of the most important MO's are given in Fig. 9, together with the excitation energies, derived from the optical absorption spectrum.

There exists a fair agreement between the calculated and the experimental energy levels. The sequence of the 3*d* MO's 17–21 is the same as that expected on the basis of EPR measurements (*64*). The high energy of the "$3\,d_{xy}$" MO accounts for the spin-pairing of the fifth 3*d* electron. In table 5 the coefficients of the 3*d* and 4*p* orbitals are listed for the MO's of interest. It can be seen that the covalency is strongly anisotropic.

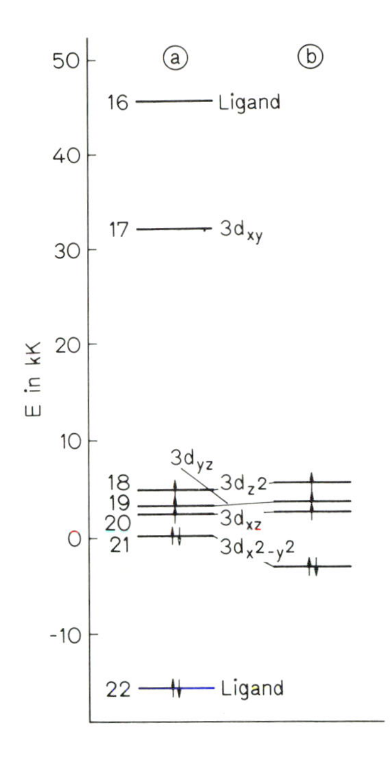

Fig. 9. Calculated relative energies (in kK) of the most important MO's (a) and spectral excitation energies derived from the electronic absorption spectrum (b) of $ClFe(Et_2dtc)_2$. The zero energy level points are taken arbitrarily. For the numbering of the MO's see ref. (*61*).

The second row of Table 6 gives the net populations of the $3d$ and $4p$ atomic orbitals. They are quite different from those predicted by the crystal field model, given in the first row.

The most striking feature is that the $3d_{xy}$ orbital, which points to the sulfur atoms, is not empty but occupied by 0.7 electronic charge. This result is not unexpected because an electron transfer from the negative dithiocarbamato ligands to the Fe(III) ion will take place via the initially empty, σ-bonding $3d_{xy}$ orbital. The interaction via the π-bonding orbitals $3d_{x^2-y^2}$, $3d_{xz}$, $3d_{yz}$ is much less important than the interaction involving the $3d_{xy}$ σ-bonding orbital. Furthermore, it is clear that the $4p$ orbitals have to be taken into account because they participate substantially in the bonding. The cal-

Table 5. MO coefficients calculated for $ClFe(Et_2dtc)_2$.

MO number	orbital type	$3d_{xy}$	$3d_{z^2}$	$3d_{yz}$	$3d_{xz}$	$3d_{x^2-y^2}$	$4p_x$	$4p_y$	$4p_z$
17	$3d_{xy}$	0.818	0	0	0	0	0	0	0
18	$3d_{z^2}$	0	0.924	0	0	−0.083	0	0	−0.205
19	$3d_{yz}$	0	0	0.941	0	0	0	0.107	0
20	$3d_{xz}$	0	0	0	0.959	0	0.098	0	0
21	$3d_{x^2-y^2}$	0	0.091	0	0	0.994	0	0	−0.024

Table 6. Net orbital populations of $ClFe(Et_2dtc)_2$ and $Fe(Et_2dtc)_3$ and the resulting QS.

	2 S + 1	Net orbital populations								QS [a]
		$3\,d_{xy}$	$3\,d_{z^2}$	$3\,d_{yz}$	$3\,d_{xz}$	$3\,d_{x^2-y^2}$	$4\,p_x$	$4\,p_y$	$4\,p_z$	
$ClFe(Et_2dtc)_2$	4 [b]	0	1	1	1	2	0	0	0	0
		0.73	1.06	1.12	1.06	2.00	0.20	0.19	0.16	2.52 ± 0.36
$Fe(Et_2dtc)_3$	2	1.68	0.77	1.66	1.68	0.75	0.13	0.13	0.14	−0.14 ± 0.02

[a]) Quadrupole splitting in mm/sec, calculated with Q = 0.21 ± 0.03 barns. Multicentre integrals have been neglected. The indicated error is due to the uncertainty in Q only.
[b]) Atomic orbital populations and resulting QS in a crystal field model.

culated QS of + 2.52 ± 0.36 mm/sec is in reasonable agreement with the observed value of + 2.76 mm/sec at liquid nitrogen temperature. From the results of this calculation it is justified to conclude that the covalency effects are primarily responsible for the observed large QS in $ClFe(Et_2dtc)_2$.

Tris(diethyldithiocarbamato)iron(III)
The QS of the six-coordinated low-spin (S = 1/2) $Fe(Et_2dtc)_3$ complex was calculated to investigate the merit of this method. If the method is useful a small QS should result because this compound is nearly octahedral. The calculations were performed with the structural parameters of $Co(Et_2dtc)_3$ (*80*) and resulted in a strong mixing of the t_{2g} and the e_g functions. From the net AO populations in Table 6 it is noticed that the π-bonding orbitals have been occupied by almost five electrons as expected, but that the σ-bonding "e_g" orbitals contain about 0.8 electronic charge each, whereas they are empty in a crystal field model. The calculated QS is small, 0.14 mm/sec, and the deviation form the observed QS (0.55 mm/sec at 77 K) may be caused by small differences in the structure of $Co(Et_2dtc)_3$ and $Fe(Et_2dtc)_3$.

In conclusion, it might be said that the method gives useful results for crystals with distinct molecules. It shows that the large QS in the five-coordinated complexes $XFe(R_2dtc)_2$ is primarily caused by covalency effects and is almost entirely due to the valence iron electrons.

Photoelectron Spectroscopy

This relatively new technique seems to be a promising one for measuring oxidation numbers of atoms. A first study of a series of transition metal dithiocarbamato complexes was published by *Frost et al.* (*189*), but these authors did not make an attempt to assign oxidation numbers. For the time being, technical difficulties as charging effects and decomposition of the samples under the influence of X-rays hinder a comparison of binding energies in atoms of different compounds. On the other hand, it is quite possible to compare atoms within one compound (e.g. in $[Cu_3(Bu_2dtc)_6]$-$[Cd_2Br_6]$ where the existence of two kinds of copper atoms clearly has been demonstrated). Another possibility to circumvent charging effects is to measure relative to an unsuspected reference. The often used C(1 s) peak of cellulose tape does not seem to be useful as reference because the tape and the glued sample may be charged differently (*190*). One probably has to search more for internal references, such as the N(1*s*) peaks of tetrabutylammonium in maleonitriledithiolato complexes and perhaps the N(1*s*) peaks in dithiocarbamates.

On the other hand, this technique provides a method to detect easily the difference between a symmetrically bonded dithiocarbamato ligand and a thiuram disulfide ligand, thus giving an indirect indication for the oxidation state of the transition metal. A complex with the former ligand should give one S(1*s*) peak whereas for a complex with the latter ligand two absorptions should be observed, due to the sulfur atoms in two different chemical environments. It should be noted, however, that an asymmetrically bonded dithiocarbamato ligand equally gives rise to two peaks for the S(1*s*) pattern.

Voltametric Studies

Many dithiocarbamato complexes have been studied with voltametric techniques. Comparison of data, however, is often difficult because different solvents and reference electrodes have been used. The solvent and the reference electrodes used in recent studies of *Martin c.s.* (*43, 68, 162,* acetone and an Ag/AgCl, 0.1 M LiCl electrode) and those in our work (*34, 37, 56, 150, 163,* CH_2Cl_2 and a saturated calomel electrode) give results sufficiently close to make a general discussion of the data possible.

The dependency of $E_{1/2}$ on the substituent R is roughly the same for all dithiocarbamato complexes. This is most clearly demonstrated with the linear plots of half-wave potentials of various electron transfers with a great variety of substituents (*43, 68*). These plots show that the order of redox stability varies from the dibenzyl compounds, which are most easy to reduce and most difficult to oxidise, to the dicyclohexyl compounds, which show just the opposite behaviour.

The Taft relation: $E_{1/2} = \rho \Sigma \sigma + x$, which was found to hold for organic compounds and some transition metal complexes can also be of use here (*37*). Phenyl compounds do not fit the relation. This is probably due to a mesomeric effect that depends on the dihedral angle between the phenyl and the NCS_2 planes. For bulky substituents deviations are also found which could be caused by widening of the CNC angle, changing the hybridisation of the N. The low values of ρ indicate that the M.O.'s involved in the electron transfer have little ligand contribution.

The redox potentials vary greatly with the nature of the metal: the electron transfer clearly regards a molecular orbital with predominant metal character (*162*). This is

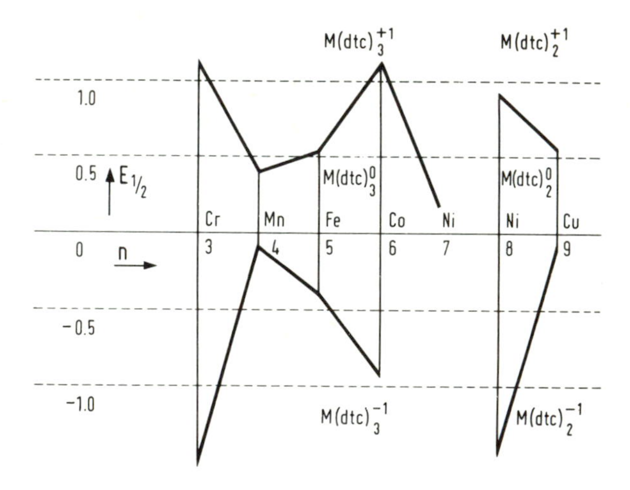

Fig. 10. Half wave potentials (at a rotating platinum electrode) vs. *d*-electron configuration for Et_2*dtc* complexes. The $E_{1/2}$ values depend upon solvent and reference electrode used (see text), but this is a minor effect as compared with the influence of the *d*-electron configuration.

in contrast with systems where the redox orbital is essentially ligand-based, e.g. in dithiolene and porphyrine complexes.

The dependency of $E_{1/2}$ on oxidation and reduction of tris(dithiocarbamato) and bis(dithiocarbamato) complexes is illustrated in Fig. 10. The great stability of the d^3 and d^6 compounds *viz.* $Cr(R_2dtc)_3$ and $Co(R_2dtc)_3$ both to oxidation and reduction is remarkable. A similar trend is present in recently published data (*191*) of the bipy complexes $M(bipy)_3^{2+}$, $M(bipy)_3$ and $M(bipy)_3^{-1}$. Maxima in redox stability are found for M = Fe(II), Cr(0) and V(– 1), respectively, which all have d^6 configuration. For the few bis(dithiocarbamato) complexes known, the stability of the d^8 Ni(II) complexes is greater than for the d^9 Cu(II) complexes.

The explanation of this peculiar dependency of $E_{1/2}$ on the electron configuration is not clear at the moment. It is improbable that this has any correlation with the well-known inertness of d^3 and d^6 octahedral complexes to ligand substitution as the voltametric data point to reversible processes, the $E_{1/2}$ values thus having thermodynamic and not kinetic relevance. The maxima in redox stability for $t_{2g}^3(^4A_{1g})$ and $t_{2g}^6(^1A_{1g})$ in O_h symmetry, and of $e_g^4 b_{1g}^2 a_{1g}^2(^1A_{1g})$ in D_{4h}, as well as the stability of the d^0 $Ti(dtc)_4$ and the d^{10} $Zn(dtc)_2$, indicate a relatively great stability for electronic states with symmetrical orbital functions. It parallels the maxima in ionisation potentials of the elements with half and completely filled subshells.

For a better comprehension, however, more systematic studies of dithiocarbamato and other complexes are necessary.

References

1. *Nyholm, R.S., Tobe, M.L.:* Adv. Inorg. Chem. Radiochem. *5*, 1 (1963).
2. *Coucouvanis, D.:* Progr. Inorg. Chem. *11*, 233 (1970).

2[a]. *Eisenberg, R.:* Progr. Inorg. Chem. *12*, 295 (1970).

3. *Pignolet, L.H.:* Topics in Current Chemistry *56*, 91 (1975).
4. *Bradley, D.C., Gitlitz, M.H.:* J. Chem. Soc. A 1152 (1969).
5. *Bradley, D.C., Rendall, I.F., Sales, K.D.:* J.C.S. Dalton 2228 (1973).
6. *Bhat, A.N., Fay, R.C., Lewis, D.F., Lindmark, A.F., Strauss, S.H.:* Inorg. Chem. *13*, 886 (1974).
7. *Colapietro, M., Vaciago, A., Bradley, D.C., Hursthouse, M.B., Rendall, I.F.:* Chem. Comm. 743 (1970).
8. *Colapietro, M., Vaciago, A., Bradley, D.C., Hursthouse, M.B., Rendall, I.F.:* J.C.S. Dalton 1052 (1972).
9. *Alyea, E.C., Ramaswamy, B.S., Bhat, A.N., Fay, R.C.:* Inorg. Nucl. Chem. Lett. *9*, 399 (1973).
10. *Que, L., Jr., Pignolet, L.H.:* Inorg. Chem. *13*, 351 (1974).
11. *Bradley, D.C., Gitlitz, M.H.:* Chem. Comm. 289 (1965).
12. *Alyea, E.C., Bradley, D.C.:* J. Chem. Soc. A 2330 (1969).
13. *Machin, D.J., Sullivan, J.F.:* J. Less-Common Met. *19*, 413 (1969).
14. *Smith, J.N., Brown, T.M.:* Inorg. Nucl. Chem. Lett. *6*, 441 (1970).
15. *Brown, T.M., Smith, J.N.:* J.C.S. Dalton 1614 (1972).
16. *Heckley, P.R., Holah, D.G.:* Inorg. Nucl. Chem. Lett. *6*, 865 (1970).
17. *Heckley, P.R., Holah, D.G., Brown, D.:* Can. J. Chem. *49*, 1151 (1971).
18. *Fackler, J.P., Jr., Holah, D.G.:* Inorg. Nucl. Chem. Lett. *2*, 251 (1966).
19. *Larkworthy, L.F., Patel, R.R.:* Inorg. Nucl. Chem. Lett. *8*, 139 (1972).
20. *Mitchell, W.J., DeArmond, M.K.:* J. Mol. Spectr. *41*, 33 (1972).
21. *Schreiner, A.F., Hauser, P.J.:* Inorg. Chem. *11*, 2706 (1972).
22. *Ortelli, J., Lacroix, R.:* Helv. Phys. Acta *41*, 1345 (1968).
23. *Price, E.R., Wasson, J.R.:* J. Inorg. Nucl. Chem. *36*, 67 (1974).
24. *Contreras, G., Cortes, H.:* J. Inorg. Nucl. Chem. *33*, 1337 (1971).
25. *Steele, D.F., Stephenson, T.A.:* Inorg. Nucl. Chem. Lett. *9*, 777 (1973).
26. *Ricard, L., Estienne, J., Weiss, R.:* Inorg. Chem. *12*, 2182 (1973).
27. *Gal, A.W., van der Ploeg, A.F.J.M., Vollenbroek, F.A., Bosman, W.:* J. Organometal. Chem. *96*, 123 (1975).
28. *Brown, D.A., Gordon, B.J., Glass, W.K., O'Daley, G.J.:* Proc. XIVth I.C.C.C. Toronto, 646 (1972).
29. *Mitchell, P.H.C., Scarle, R.D.:* J.C.S. Dalton 110 (1975).
30. *Jowitt, R.N., Mitchell, P.H.C.:* Inorg. Nucl. Chem. Lett. *4*, 39 (1968).
31. *Bradley, D.C., Chisholm, M.H.:* J. Chem. Soc. A 2741 (1971).
32. *Nieuwpoort, A., Moonen, J.H.E., Cras, J.A.:* Rec. Trav. Chim. *92*, 1068 (1973).
33. *Nieuwpoort, A.:* J. Less-Common Met. *36*, 271 (1974).
34. *Vàràdi, Z.B., Nieuwpoort, A.:* Inorg. Nucl. Chem. Lett. *10*, 801 (1974).
35. *van der Aalsvoort, J.G.M., Beurskens, P.T.:* Cryst. Struct. Comm. *3*, 653 (1974).
36. *Wijnhoven, J.G.:* Cryst. Struct. Comm. *2*, 637 (1973).
37. *Nieuwpoort, A.:* Thesis Nijmegen (1975).
38. *Newton, W.E., Bravard, D.C., McDonald, J.W.:* Inorg. Nucl. Chem. Lett. *11*, 553 (1975).
39. *Bishop, M.W., Chatt, J., Dilworth, J.R.:* J. Organometal. Chem. *73*, C59 (1974).
40. *Kirmse, R., Hoyer, E.:* Z. anorg. allg. Chem. *398*, 136 (1973).
41. *Brown, D., Holah, D.G., Rickard, C.E.F.:* J. Chem. Soc. A423, 786 (1970).
42. *Cambi, L., Cagnasso, A.:* Atti Acad. Lincei *14*, 71 (1931); Chem. Abstr. *26*, 2172[4] (1932).
43. *Hendrickson, A.R., Martin, R.L., Rohde, N.M.:* Inorg. Chem. *13*, 1933 (1974).

44. *Lahiry, S., Anand, V.K.:* Chem. Comm. 1111 (1971).
45. *Holah, D.G., Murphy, C.N.:* Can. J. Chem. *49*, 2726 (1971).
46. *Delepine, M.:* Compt. Rend. *144*, 1125 (1907).
47. *Healy, P.C., White, A.H.:* J.C.S. Dalton 1883 (1972).
48. *Prabhakaran, C.P., Patel, C.C.:* Ind. J. Chem. *7*, 1257 (1969).
49. *Dingle, R.:* Acta Chem. Scand. *20*, 33 (1966).
50. *Figgis, B.N., Toogood, G.E.:* J.C.S. Dalton 2177 (1972).
51. *Pasek, E.A., Straub, D.K.:* Inorg. Chem. *11*, 259 (1972).
52. *Golding, R.M., Harris, C.M., Jessop, K.J., Tennant, W.C.:* Aust. J. Chem. *25*, 2567 (1972).
53. *Saleh, R.Y., Straub, D.K.:* Inorg. Chem. *13*, 3017 (1974).
54. *Brown, K.L., Golding, R.M., Healy, P.C., Jessop, K.J., Tennant, W.C.:* Aust. J. Chem. *27*, 2075 (1974).
55. *Brown, K.L.:* Cryst. Struct. Comm. *3*, 493 (1974).
56. *Nieuwpoort, A.:* to be published.
57. *Rowbottom, J.F., Wilkinson, G.:* J.C.S. Dalton 826 (1972).
58. *Colton, R., Levitus, R., Wilkinson, G.:* J. Chem. Soc. 5275 (1960).
59. *Rowbottom, J.F., Wilkinson, G.:* Inorg. Nucl. Chem. Lett. *9*, 675 (1973).
60. *Rowbottom, J.F.:* personal communication.
61. *de Vries, J.L.K.F.:* Thesis Nijmegen (1972).
62. *Hoskins, B.F., Kelly, B.P.:* Chem. Comm. 1517 (1968).
63. *Healy, P.C., White, A.H.:* J.C.S. Dalton 1165 (1972).
64. *Martin, R.L., White, A.H.:* Inorg. Chem. *6*, 712 (1967).
65. *Hoskins, B.F., Martin, R.L., White, A.H.:* Nature *211*, 627 (1966).
66. *Davies, G.R., Jarvis, J.A.J., Kilburn, B.T., Mais, R.H.B., Owston, P.G.:* J. Chem. Soc. A 1275 (1970).
67. *de Vries, J.L.K.F., Trooster, J.M., de Boer, E.:* Inorg. Chem. *10*, 81 (1971).
68. *Chant, R.L., Hendrickson, A.R., Martin, R.L., Rohde, N.M.:* Inorg. Chem. *14*, 1894 (1975).
69. *Martin, R.L., Rohde, N.M., Robertson, G.B., Taylor, D.:* J. Am. Chem. Soc. *96*, 3647 (1974).
70. *Leipoldt, J.G., Coppens, P.:* Inorg. Chem. *12*, 2269 (1973).
71. *Cambi, L., Malatesta, L.:* Rend. In Lombardo Sci. Lettre A 7', 118 (1938); Chem. Abstr. *34*, 3201 (1940).
72. *Malatesta, L.:* Gazz. Chim. Ital. *68*, 195 (1938).
73. *Domenicano, A., Vaciago, A., Zambonelli, L., Loader, P.L., Venanzi, L.M.:* Chem. Comm. 476 (1966).
74. *Pignolet, L.H.:* Inorg. Chem. *13*, 2051 (1974).
75. *Patterson, G.S., Holm, R.H.:* Inorg. Chem. 2285 (1972).
76. *Gahan, L.R., O'Connor, M.J.:* J.C.S. Chem. Comm. 68 (1974).
77. *Pignolet, L.H., Mattson, B.M.:* J.C.S. Chem. Comm. 49 (1975).
78. *Compin, L.M.:* Bull. Soc. Chim. France *27*, 464 (1920).
79. *Marcotrigiano, G., Pellacani, G.C., Preti, C.:* J. Inorg. Nucl. Chem. *36*, 3709 (1974).
80. *Merlino, S.:* Acta Cryst. *B24*, 1441 (1968).
81. *Hendrickson, A.R., Martin, R.L.:* J.C.S. Chem. Comm. 873 (1974).
82. *Marcotrigiano, G., Pellacani, G.C., Preti, C., Tosi, G.:* Bull. Chem. Soc. Japan *48*, 1018 (1975).
83. *de Croon, M.H.J.M., van Gaal, H.L.M., van der Ent, A.:* Inorg. Nucl. Chem. Lett. *10*, 1081 (1974).
84. *Connelly, N.G., Green, M., Kuc, T.A.:* J.C.S. Chem. Comm. 542 (1974).
85. *Connelly, N.G.:* personal communication.
86. *Mitchell, R.W., Ruddick, J.O., Wilkinson, G.:* J. Chem. Soc. A 3224 (1971).
87. *Sceney, G.C., Magee, R.J.:* Inorg. Nucl. Chem. Lett. *9*, 251 (1973).
88. *Capacchi, L., Nardelli, M., Villa, A.:* Chem. Comm. 441 (1966).
89. *Gasparri, G.F., Nardelli, M., Villa, A.:* Acta Cryst. *23*, 348 (1967).
90. *Newman, P.W.G., White, A.H.:* J.C.S. Dalton 1460 (1972).

91. *Bonamico, M., Dessy, G., Mariani, C., Vaciago, A., Zambonelli, L.:* Acta Cryst. *19*, 619 (1965).
92. *Peyronel, G., Pignedoli, A.:* Acta Cryst. *23*, 399 (1967).
93. *Tomlinson, A.A., Furlani, C.:* Inorg. Chim. Acta *3*, 487 (1969).
94. *Lachenal, D.:* Inorg. Nucl. Chem. Lett. *11*, 101 (1975).
95. *Solozjenkin, P.M., Kopitsja, N.:* Dokl. Akad. Nauk. Tadzh. S.S.R. *12*, 30 (1969).
96. *van der Linden, J.G.M.:* to be published.
97. *Willemse, J., Rouwette, P.H.F.M., Cras, J.A.:* Inorg. Nucl. Chem. Lett. *8*, 389 (1972).
98. *Brinkhoff, H.C.:* Rec. Trav. Chim. *90*, 377 (1970).
99. *Brinkhoff, H.C.:* Thesis Nijmegen (1970).
100. *Brinkhoff, H.C., Cras, J.A., Steggerda, J.J., Willemse, J.:* Rec. Trav. Chim. *88*, 633 (1969).
101. *Fackler, J.P., Jr., Avdeef, A., Fischer, R.G.; Jr.:* J. Am. Chem. Soc. *95*, 774 (1973).
102. *Willemse, J.:* Thesis Nijmegen (1974).
103. *Beurskens, P.T., Cras, J.A.:* J. Cryst. Mol. Struct. *1*, 63 (1971).
104. *Avdeef, A., Fackler, J.P., Jr., Fischer, J.G., Jr.:* J. Am. Chem. Soc. *92*, 6972 (1970).
105. *Noordik, J.H., Smits, J.M.M.:* Cryst. Struct. Comm. *3*, 253 (1974).
106. *Nigo, Y., Masuda, I., Shinra, K.:* Chem. Comm. 476 (1970).
107. *Jensen, K.A., Krishnan, V.:* Acta Chem. Scand. *24*, 1088 (1970).
108. *Willemse, J., Cras, J.A.:* J. Inorg. Nucl. Chem., in press.
109. *Hieber, W., Brück, R.:* Z. anorg. allg. Chem. *269*, 13 (1952).
110. *Fackler, J.P., Jr., Coucouvanis, D.:* J. Am. Chem. Soc. *89*, 1745 (1967).
111. *Sceney, G., Magee, R.J.:* Inorg. Nucl. Chem. Lett. *10*, 323 (1974).
112. *Beurskens, P.T., Cras, J.A., Hummelink, Th.W., Noordik, J.H.:* J. Cryst. Mol. Struct. *1*, 253 (1971).
113. *Amanov, A.Z., Kukina, Ġ.A., Porai-Koshits, M.A.:* Zh. Struct. Khim. *8*, 174 (1967).
114. *Briscoe, G.B., Humphries, S.:* Talanta *16*, 1403 (1969).
115. *Dahl, L.F., Martell, C., Wampler, D.L.:* J. Am. Chem. Soc. *83*, 1762 (1961).
116. *Willemse, J., Cras, J.A., Wijnhoven, J.G., Beurskens, P.T.:* Rec. Trav. Chim. *92*, 1199 (1973).
117. *Willemse, J., Cras, J.A.:* Rec. Trav. Chim. *91*, 1309 (1972).
118. *Bosman, W.P.J.H., Wijnhoven, J.G., Beurskens, P.T.:* to be published.
119. *Willemse, J., Cras, J.A., van de Leemput, P.J.H.A.M.:* Inorg. Nucl. Chem. Lett. *12*, 255 (1976).
120. *Beurskens, P.T., Bosman, W.P.J.H., Cras, J.A.:* J. Cryst. Mol. Struct. *2*, 183 (1972).
121. *Åkerström, S.:* Arkiv Kemi *14*, 387 (1959).
122. *Åkerström, S.:* Arkiv Kemi *14*, 403 (1959).
123. *Hesse, R.:* Arkiv Kemi *20*, 481 (1963).
124. *Tamminen, V., Hjelt, E.:* Suomen Kemistilehti *B23*, 39 (1950).
125. *Cambi, L., Coriselli, C.:* Gazz. Chim. Ital. *66*, 779 (1936).
126. *Fredga, A.:* Rec. Trav. Chim. *69*, 420 (1950).
127. *Janssen, M.J.:* Rec. Trav. Chim. *75*, 1141 (1956).
128. *Janssen, M.J.:* Rec. Trav. Chim. *76*, 827 (1957).
129. *Vänngard, T., Åkerström, S.:* Nature *184*, 183 (1959).
130. *Weeks, M.J., Fackler, J.P., Jr.:* Inorg. Chem. 7, 2548 (1968).
131. *Gregson, A.K., Mitra, S.:* J. Chem. Phys. *49*, 3696 (1968).
132. *de Villa, J.F., Chatfield, D.A., Bursey, M., Hatfield, W.E.:* Inorg. Chim. Acta *6*, 332 (1972).
133. *Pilbrow, J.R., Toy, A.D., Smith, T.D.:* J. Chem. Soc. A 1029 (1969).
134. *Bonamico, M., Dessy, G., Mugnoli, A., Vaciago, A., Zambonelli, L.:* Acta Cryst. *19*, 886 (1965).
135. *Regenass, W., Fallab, S., Erlenmeyer, H.:* Helv. Chim. Acta *38*, 1448 (1955).
136. *Taylor, D.:* personal communication.
137. *Beurskens, P.T., Cras, J.A., Steggerda, J.J.:* Inorg. Chem. 7, 810 (1968).
138. *Wijnhoven, J.G., van der Hark, Th.E.M., Beurskens, P.T.:* J. Cryst. Mol. Struct. *2*, 189 (1972).

139. *Cras, J.A., Willemse, J., Gal, A.W., Hummelink-Peters, B.G.M.C.:* Rec. Trav. Chim. *92*, 641 (1973).
140. *van de Leemput, P.J.H.A.M., Cras, J.A., Willemse, J.:* Rec. Trav. Chim. *95*, 191 (1976).
141. *Cras, J.A.:* Proc. XVIth I.C.C.C. Dublin (1974).
142. *van de Leemput, P.J.H.A.M., Willemse, J., Cras, J.A.:* Rec. Trav. Chim. *95*, 53 (1976).
143. *Golding, R.M., Rae, A.D., Ralph, B.J., Sulligoi, L.:* Inorg. Chem. *13*, 2499 (1974).
144. *Hesse, R., Nilson, L.:* Acta Chem. Scand. *23*, 825 (1969).
145. *Bergendahl, T.J., Bergendahl, E.M.:* Inorg. Chem. *11*, 638 (1972).
146. *Hesse, R.:* in "Advances in the Chemistry of Coordination Compounds", S. Kirschner, Ed., MacMillan, New York, 324 (1961).
147. *Sutton, L.E.*, Ed.: Chem. Soc., Spec. Publ. *No 11*, S3 (1958).
148. *Farrell, F.J., Spiro, T.G.:* Inorg. Chem. *10*, 1606 (1971).
149. *van Rens, J.G.M.:* Thesis Nijmegen (1974).
150. *van der Linden, J.G.M.:* Thesis Nijmegen (1972).
151. *Beurskens, P.T., Cras, J.A., van der Linden, J.G.M.:* Inorg. Chem. *9*, 475 (1970).
152. *Beurskens, P.T., Cras, J.A., Hummelink, Th.W., van der Linden, J.G.M.:* Rec. Trav. Chim. *89*, 984 (1970).
153. *van der Linden, J.G.M.:* Rec. Trav. Chim. *90*, 1027 (1971).
154. *Noordik, J.H.:* Cryst. Struct. Comm. *2*, 81 (1973).
155. *Beurskens, P.T., Blaauw, H.J.A., Cras, J.A., Steggerda, J.J.:* Inorg. Chem. *7*, 805 (1968).
156. *Cras, J.A., Noordik, J.H., Beurskens, P.T., Verhoeven, A.M.:* J. Cryst. Mol. Struct. *1*, 155 (1971).
157. *Noordik, J.H.:* Thesis Nijmegen (1971).
158. *Gal, A.W., Beurskens, G., Cras, J.A., Beurskens, P.T., Willemse, J.:* Rec. Trav. Chim. *95*, 157 (1976).
159. *Blaauw, H.J.A., Nivard, R.J.F., van der Kerk, G.J.M.:* J. Organometal. Chem. *2*, 236 (1964).
160. *Blaauw, H.J.A.:* Thesis Nijmegen (1965).
161. *Beurskens, P.T., Cras, J.A., Noordik, J.H., Spruijt, A.M.:* J. Cryst. Mol. Struct. *1*, 93 (1971).
162. *Chant, R., Hendrickson, A.R., Martin, R.L., Rohde, N.M.:* Aust. J. Chem. *26*, 2533 (1973).
163. *van der Linden, J.G.M., van de Roer, H.G.J.:* Inorg. Chim. Acta *5*, 254 (1971).
164. *Martin, R.L., White, A.H.:* Transition Metal Chem. *4*, 113 (1968).
165. *Shopov, D., Yordanov, N.D.:* Inorg. Chem. *9*, 1943 (1970).
166. *Yordanov, N.D., Shopov, D.:* Inorg. Chim. Acta *5*, 679 (1971).
167. *Wayland, B.B., Wisniewski, M.D.:* Chem. Comm. 1025 (1971).
168. *Herring, F.G., Tapping, R.L.:* Can. J. Chem. *52*, 4016 (1974).
169. *Toy, A.D., Chaston, S.H.H., Pilbrow, J.R., Smith, T.D.:* Inorg. Chem. *10*, 2219 (1971).
170. *Golding, R.M., Sinn, E., Tennant, W.C.:* J. Chem. Phys. *56*, 5296 (1972).
171. *DeSimone, R.E.:* J. Am. Chem. Soc. *95*, 6238 (1973).
172. *Cowsik, R.K., Rangarajan, G., Srinivasan, R.:* Chem. Phys. Lett. *8*, 136 (1970).
173. *van Rens, J.G.M., van der Drift, E., de Boer, E.:* Chem. Phys. Lett. *14*, 113 (1972).
174. *van Rens, J.G.M., de Boer, E.:* Chem. Phys. Lett. *31*, 377 (1975).
175. *van Willigen, H., van Rens, J.G.M.:* Chem. Phys. Lett. *2*, 283 (1968).
176. *van Rens, J.G.M., Viegers, M.P.A., de Boer, E.:* Chem. Phys. Lett. *28*, 104 (1974).
177. *Schlupp, R.L., Maki, A.H.:* Inorg. Chem. *13*, 44 (1974).
178. *Keijzers, C.P., de Vries, H.J.M., van der Avoird, A.:* Inorg. Chem. *11*, 1338 (1972).
179. *Keijzers, C.P.:* Thesis Nijmegen (1974).
180. *Keijzers, C.P., de Boer, E.:* J. Chem. Phys. *57*, 1277 (1972).
181. *Keijzers, C.P., de Boer, E.:* Mol. Phys. *29*, 1007 (1975).
182. *de Vries, J.L.K.F., Keijzers, C.P., de Boer, E.:* Inorg. Chem. *11*, 1343 (1972).
183. *Mulliken, R.S.:* J. Chem. Phys. *23*, 1833 (1955).
184. *Frank, E., Abeledo, C.R.:* J. Inorg. Nucl. Chem. *31*, 989 (1969).
185. *de Vries, J.L.K.F., Trooster, J.M., de Boer, E.:* Inorg. Chem. *12*, 2730 (1973).
186. *Epstein, L.M.; Straub, D.K.:* Inorg. Chem. *8*, 784 (1969).

187. *Wickmann, H. H., Trozzolo, A. M.:* Inorg. Chem. *7*, 63 (1968).
188. *Epstein, L. M., Straub, D. K.:* Inorg. Chem. *8*, 560 (1969).
189. *Frost, D. C., McDowell, C. A., Tapping, R. L.:* J. Electron Spectrosc. *7*, 297 (1975).
190. *Trooster, J. M.:* personal communication.
191. *Saji, T., Aoyagui, S.:* J. Electroanal. Chem. *63*, 405 (1975).
192. *Hill, D. M., Larkworthy, L. F., O'Donoghue, M. W.:* J.C.S. Dalton 1726 (1975).
193. *McCleverty, J. A., Morrison, N. J.:* J.C.S. Chem. Comm. 108 (1974).
194. *van de Leemput, P. J. H. A. M., Cras, J. A., Willemse, J.:* unpublished work.
195. *Hendrickson, A. R., Martin, R. L., Taylor, D.:* J.C.S. Chem. Comm. 843 (1975).

Author-Index Volume 1–28

CH. K. JØRGENSEN

Oxidation Numbers and Oxidation States

VII, 291 pages. 1969
(Molekülverbindungen und Koordinationsverbindungen in Einzeldarstellungen)

Contents: Introduction. – Formal Oxidation Numbers. – Configurations in Atomic Spectroscopy. – Characteristics of Transition Group Ions. – Internal Transitions in Partly Filled Shells. – Inter-Shell Transitions. – Electron Transfer Spectra and Collectively Oxidized Ligands. – Oxidation States in Metals and Black Semi-Conductors. – Closed-Shell Systems, Hydrides and Back-Bonding. – Homopolar Bonds and Catenation. – Quanticule Oxidation States. – Taxological Quantum Chemistry.

For most chemists, formal oxidation numbers are a tool for writing correct reaction schemes more conveniently. However, recent progress in absorption spectroscopy of transition group complexes and the application of group theory and molecular orbital theory to chromophores consisting of a central atom and a number of adjacent atoms belonging to the ligands has made it possible to connect the classification of oxidation states with spectroscopic and magnetochemical results. Consequently, the apparent discrepancy between the integral number of electrons in the partly filled shell and the fractional atomic charges can be clarified in a satisfactory and rather unexpected fashion. However, under certain circumstances, the ligands are not innocent and do not allow the determination of oxidation states. The book treats many detailed problems, such as the phenomenological baricenter polynomial for electron configurations in monatomic entities, spin-pairing energy, the Madelung potential and the stabilization of definite oxidation states, the nephelauxetic effect, bonding in semiconductors, the question of back-bonding in carbonyl, olefin and hydride complexes, etc.

M. MEHRING

High Resolution NMR Spectroscopy in Solids

104 figures. Approx. 240 pages. 1976
(NMR – Basic Principles and Progress, Vol. 11)

Contents: Introduction. – Nuclear Spin Interactions in Solids. – Multiple-Pulse NMR Experiments. – Double Resonance Experiments. – Magnetic Shielding Tensor. – Spin-Lattice Relaxation in Line Narrowing Experiments. – Appendix.

This book comprises the first and only comprehensive representation of high resolution NMR techniques in solids up to date. It deals with a great variety of applications, reaching from metals to organic and macromolecular solids.

Springer-Verlag
Berlin Heidelberg New York

Electrons in Fluids

The Nature of Metal-Ammonia Solutions
Editors: J. Jortner, N.R. Kestner
271 figures, 59 tables. XII, 493 pages. 1973
With contributions by numerous experts

Contents: Theory of Electrons in Polar Fluids. – Metal-Ammonia Solutions: The Dilute Region. – Metal Solutions in Amines and Ethers. – Ultrafast Optical Processes. – Metal-Ammonia Solutions: Transition Range. – The Electronic Structures of Disordered Materials. – Concentrated M-NH_3-Solutions: A Review. – Strange Magnetic Behavior and Phase Relations of Metal-Ammonia Compounds. – Metallic Vapors. – Mobility Studies of Excess Electrons in Nonpolar Hydrocarbons. – Optical Absorption Spectrum of the Solvated Electron in Ethers and in Binary Liquid Systems. – Subject Index.
Color Plates.

This full and up-to-date account of the chemical and physical properties of electrons in polar, non-polar, and dense fluids includes contributions from both theoretical and experimental chemists and physicists, thus clearly indicating the interdisciplinary nature of this field.

**Springer-Verlag
Berlin
Heidelberg
New York**